Second Edition

Essence of Linear Algebra

Kim, Yang-Gon
Seo, Gyeong-Sig

KYOWOOSA

ESSENCE OF LINEAR ALGEBRA

(second edition)

Kim, Yang-Gon
Seo, Gyeong-Sig

Typeset by $\mathcal{AMS}$-TEX

Yanggon Kim and Gyeongsig Seo
Department of Mathematics and
Institute of Pure and Applied Mathematics
Chonbuk National University
Chonju, Chonbuk 561-756, Korea

2000 AMS Subject Classification : 15-00, 15-01
Library of Congress Cataloging in Publication Data

Kim, Yanggon and Seo, Gyeongsig

Essence of Linear Algebra (Undergraduate text in mathematics)

This volume was directly printed from the copy prepared by the authors using $\mathcal{A}\mathcal{M}\mathcal{S}$-TEX.

PREFACE

Some nice text for one year undergraduate course has been intended by the authors after many years experience of teaching of Linear algebra. As is well-known, Linear algebra is very important not only in every discipline of mathematics but also in other branches of study such as physics, engineering, economics etc. Of course there are so many stuffs relating to Linear algebra. Among those only essential ones, however, are included in this book.

This book consists of 10 chapters with each chapter, as it were, very important as well as very essential in its own right. If there is any defect in this book, it shall be removed of this little by little and the authors would like to improve this book at the same time. In closing this prelude the authors should acknowledge all who helped us write this book secretively or positively. In particular S.H.Koh who is the authors' student as well as a doctor in mathematics, suffered the hard and boring work of typing the handwriting. So many thanks are due to him. In addition the authors cannot help thanking their families who have sustained them all the way.

June, 2011

Kim, Yang-Gon

Seo, Gyeong-Sig

CONTENTS

CHAPTER 1. BASIS

Vector space and its basis is very important in every branch of mathematics. Such stuffs are defined usually over certain fields. So we introduce the notion of field at first and then explain basis explicitly and centrally in this chapter .

But trivial concepts or terminologies shall be used freely without explicit definitions now and then.

§ 1-1 Fields

First, we should make clear what the field is.

Definition(1.1.1). Let F be a nonempty set with two operations $+, \times$ called *addition* and *multiplication* respectively. F is said to be a *field* if the following properties are satisfied :

(i) $\forall a, b \in F$, $a + b = b + a \in F$ and $a \times b = b \times a \in F$. The latter is written simply as $a \cdot b = b \cdot a$ or $ab = ba \in F$ if there is no confusion.

(ii) $\forall a, b, c \in F$, $(a + b) + c = a + (b + c)$ and $(ab)c = a(bc)$.

(iii) $\forall a, b, c \in F$, $(a + b)c = ac + bc$.

(iv) $\exists 0, 1 \in F$ such that (abbreviation s.t.) $\forall a \in F$, $a + 0 = a$ and $a \cdot 1 = a$.

Typeset by $\mathcal{AMS}$-TEX

(v) $\forall a \in F, \exists -a \in F$ s.t. $a+(-a)=0$. $\forall a \in F^{\times} := F-\{0\}$, $\exists a^{-1} \in F$ s.t. $a \cdot a^{-1} = 1$.

Definition(1.1.2). If there is a prime number p s.t. $pf = 0, \forall f \in F$, we say p is the *characteristic* of F. Otherwise we say F has the characteristic 0.

As we are well aware, a set is naively defined to be a collection of certain objects. We call its member an *element* of the set. We say that B is a *subset* of A if each element of B becomes an element of A. B is a *proper subset* of A if B is a subset of A and $B \neq A$. We usually write $B \subset A$ as a subset notation and $B \subsetneq A$ as a proper subset notation.

In this case we may say B is contained in A. For any sets A, B, we denote the union of A and B by $A \cup B$ and the intersection of A and B by $A \cap B$. For a set X, X^m denotes the set of m-tuples of elements of X, called the *m-times Cartesian product* of X.

Example(1.1.3). Usually $\mathbb{Q}, \mathbb{R}$ and $\mathbb{C}$ denote the set of rational numbers, real numbers, and complex numbers respectively ; these are all intimate examples of field. As a nonexample of a field we may take $\mathbb{Z}$ which denotes the well known set of all integers as well as $\mathbb{N}$, the natural numbers.

Definition(1.1.4). Elements of a field F are usually called *scalars* or *numbers* otherwise. If F' is a subset of a field F and itself is a field, F' is called a *subfield* of F.

Remark(1.1.5). We shall deal with those fields only with the characteristics making denominators nonzero in our discussion.

Exercises of § 1-1

[1] Show that $\mathbb{Q}$ is a subfield of $\mathbb{R}$ and $\mathbb{R}$ is a subfield of $\mathbb{C}$.

[2] Show that 0 and 1 are unique in any field F.

[3] May it happen that $0 = 1$ in a field F ?

[4] Besides $\mathbb{Q}, \mathbb{R}$ and $\mathbb{C}$, take an example of a field.

[5] Why can't we think of 0^{-1} ?

§ 1-2 Vector space

Vector space is usually defined over a field. It is defined explicitly in what follows. Sometimes we omit "over a field" when it is made clear in the context.

Definition(1.2.1). Let V be a nonempty set in which the following axioms are satisfied; we then call V a *vector space* over a field F:

(i) $\forall v_1, v_2 \in V,\ v_1 + v_2 = v_2 + v_1 \in V$ (addition of vectors).

(ii) $\forall f \in F, \forall v \in V,\ fv \in V$ (multiplication by scalars).

(iii) $\forall v \in V, 1 \cdot v = v$.

(iv) $\forall v_1, v_2, v_3, \in V,\ v_1 + (v_2 + v_3) = (v_1 + v_2) + v_3$.

(v) $\exists$ identity element $\mathbb{O}$ with respect to addition, i.e., $\forall v \in V,\ v + \mathbb{O} = \mathbb{O} + v = v$.

(vi) $\exists$ inverse element $-v$ of $\forall v \in V$ with respect to addition, i.e., $v + (-v) = -v + v = \mathbb{O}$.

(vii) $\forall f \in F,\ \forall v_1, v_2 \in V,\ f(v_1 + v_2) = f(v_1) + f(v_2)$.

(viii) $\forall f_1, f_2 \in F,\ \forall v \in V,\ (f_1 + f_2)v = f_1 v + f_2 v$.

(ix) $\forall f_1, f_2 \in F,\ \forall v \in V,\ (f_1 f_2)v = f_1(f_2 v)$.

Definition(1.2.2). A subset S of a vector space V over a field F is called a *subspace* of V in case that:

$\forall s_1, s_2 \in S,\ s_1 + s_2 \in S;\ \forall f \in F,\ \forall s \in S,\ fs \in S$.

Evidently S itself is a vector space over F.

Definition(1.2.3). Let $f_1, \cdots, f_\ell$ be any scalars in F and v_1,

$\cdots, v_\ell$ be any vectors of V; by a *linear combination* of $v_1, \cdots, v_\ell$ we mean $\sum_{i=1}^{\ell} f_i v_i$ which is also a vector of V.

Proposition(1.2.4). *All linear combinations of* $v_1, \cdots, v_\ell \in V$ *form a subspace of* V. *We call such a subspace a linear span of* $v_1, \cdots, v_\ell$.

Proof. For any scalars $(f_1, \cdots, f_\ell), (f_1', \cdots, f_\ell') \in F^\ell$, we have $\sum_{i=1}^{\ell} f_i v_i + \sum_{i=1}^{\ell} f_i' v_i = \sum_{i=1}^{\ell} (f_i + f_i') v_i$ which is again a linear combination of $v_i's$. Next for any scalar $f \in F$, we obtain $f(\sum_{i=1}^{\ell} f_i v_i) = \sum_{i=1}^{\ell} (f f_i) v_i$ which is also such a linear combination.

Definition(1.2.5). F^ℓ is called the set of all *ℓ-tuples* of elements of F. If $v = (f_1, \cdots, f_\ell) \in F^\ell$, then each f_i is called a *coordinate* or *component* of v.

We easily see that F^ℓ becomes a vector space with componentwise addition and componentwise scalar multiplication, i.e., $\forall f \in F,\ \forall (f_1, \cdots, f_\ell), (f_1', \cdots, f_\ell') \in F^\ell$,
we designate:

$$(f_1, \cdots, f_\ell) + (f_1', \cdots, f_\ell') := (f_1 + f_1', \cdots, f_\ell + f_\ell') \in F^\ell,$$
$$f(f_1, \cdots, f_\ell) := (f f_1, \cdots, f f_\ell) \in F^\ell.$$

Definition(1.2.6). In what follows, we frequently use the so-called usual *scalar product* or *dot product* designated by $v \cdot v' = \sum f_i f_i' \in F,\ \forall v = (f_1, \cdots, f_\ell),\ v' = (f_1', \cdots, f_\ell') \in F^\ell$.

Proposition(1.2.7). *The usual scalar product on the vector space F^ℓ possesses the following properties:*

(i) $\forall v, v' \in F^\ell,\ v \cdot v' = v' \cdot v;$

(ii) $\forall v, v', v'' \in F^\ell,\ v \cdot (v' + v'') = (v' + v'') \cdot v = v \cdot v' + v \cdot v'' = v' \cdot v + v'' \cdot v;$

(iii) $\forall v, v' \in F^\ell,\ \forall f \in F,\ (fv) \cdot v' = f(v \cdot v') = v \cdot (fv').$

Proof. (i) Since F is a field, $f_i f_i' = f_i' f_i\ \forall f_i, f_i' \in F$. So (i) is evident. Also (ii) is obvious because of property (iii) of a field in the definition(1.1.1).

Finally (iii) holds because of property (i) and (ii) of definition(1.1.1).

Definition(1.2.8). For any two vectors $v, w \in F^\ell$ with $v \cdot w = 0$, we say that v and w are *orthogonal* or *perpendicular* with respect to the usual scalar product on F^ℓ. Sometimes we denote such a relation by $v \perp w$.

Example(1.2.9). For a fixed element $v \in F^\ell$, $S := \{w \in F^\ell \mid v \cdot w = 0\}$ just becomes a subspace of the vector space F^ℓ. Its easy proof is left to readers as an exercise.

Definition(1.2.10). If a mapping has a field as its codomain, we call it a *function* usually. In particular F-valued function is obviously defined.

Example(1.2.11). Suppose that V is the collection of all functions from a set E into a field F. If $\forall \tilde{f_1}, \tilde{f_2} \in V$, $\forall e \in E, \forall f \in F$, we define $(\tilde{f}_1 + \tilde{f}_2)(e) := \tilde{f}_1(e) + \tilde{f}_2(e)$ and $(f\tilde{f}_1)(e) = f(\tilde{f}_1(e))$, then we may easily see that V is rendered a vector space over F. Such a vector space is called a *function space* in particular in the literature.

Exercises of § 1-2

[1] Show that $\forall v \in V$ (a vector space over F), $0v = \mathbb{O}$ and $\forall f \in F,\ f\mathbb{O} = \mathbb{O}$.

[2] If $\mathbb{O} \neq fv = f'v'$ for some $f, f' \in F$ and for some $v, v' \in V$, verify that

$$\begin{cases} f = f' & \text{in case } v = v', \\ v = v' & \text{in case } f = f'. \end{cases}$$

[3] Prove that S in example 1.2.9 becomes a subspace of F^ℓ.

[4] Prove that V in example 1.2.11 forms a vector space.

[5] Let $V := \{x + y\sqrt{2} \mid \forall x, y \in \mathbb{Q}\}$.

(i) Show that V is a field containing $\mathbb{Q}$.

(ii) Show that V becomes a vector space over V itself and also over $\mathbb{Q}$.

[6] Verify that any field forms a vector space itself.

§ 1-3 Definition of Basis

Needless to say, the notion of basis is central amid those relating to vector space. Prior to this, we have to undergo linear independency which seems indigenous and picky early in linear algebra.

Definition(1.3.1). Suppose that V is a vector space over F and $v_1, \cdots, v_\ell$ are vectors in V. By *linear independency* of $v_1, \cdots, v_\ell$, we mean whenever $\sum_{i=1}^{\ell} f_i v_i = \mathbb{O}$ for some $f_i' s \in F$, all f_i are zero. Otherwise we utter linear dependency of $v_i' s$. Namely if $\sum_{i=1}^{\ell} f_i v_i = \mathbb{O}$ for some nonzero $f_i \in F$, we say that $v_1, \cdots, v_\ell$ are *linearly dependent* (over F).

Example(1.3.2). Let $U_i' s$ be standard unit vectors in F^ℓ, i.e., $U_i := (0, \cdots, 0, 1, 0, \cdots, 0)$ with i-th component 1 and others zero. If $\ell \geq k$, then $U_1, \cdots, U_k$ are linearly independent vectors of F^ℓ. We may actually verify this. If $\sum_{i=1}^{k} f_i U_i = (f_1, \cdots, f_k, 0, \cdots, 0) = \mathbb{O}$, then $f_1 = \cdots = f_k = 0$. Hence $U_1, \cdots, U_k$ are linearly independent.

Furthermore any nonempty subset of $\{U_1, \cdots, U_\ell\}$ also becomes a set of linearly independent vectors.

Example(1.3.3). For any vector space V over F a finite subset of V including $\mathbb{O}$ must be a linearly dependent set.

For, consider a set $\{\mathbb{O}, v_1, v_2, \cdots, v_\ell\} \subset V$; we then have a relation $1 \cdot \mathbb{O} + \sum_{i=1}^{\ell} 0 \cdot v_i = \mathbb{O}$ which ensures the linear dependency of the set.

Definition(1.3.4). Let S be a nonempty subset of a vector space V. The set of all finite linear combinations of elements of S forms a subspace V_S of V evidently. In this case we say that S *generates* V_S or all elements of S *generate* V_S. Of course $V_S = V$ may happen for a suitable S, e.g., for $S = V$. Suppose further that any finite number of elements of S are linearly independent; we then say that S is a *basis* of V_S. We also say that V_S is *finite dimensional* if the number of all elements of a basis S is finite; otherwise we call it *infinite dimensional.* The cardinality of S is called the *dimension* of V_S and is written by $Dim_F V_S$, which is a well-defined concept due to proposition (1.3.10) below.

Example(1.3.5). Standard unit vectors in F^ℓ, i.e., $U_1, \cdots, U_\ell$ forms a basis of the vector space F^ℓ. So its dimension $Dim_F F^\ell = \ell$.

Example(1.3.6). The vector space V generated by two functions $sinx, cosx$ has $\{sinx, cosx\}$ as a basis of V. For, suppose that $r_1 sinx + r_2 cosx = 0$; differentiating both sides of this equation, we have $r_1 cosx - r_2 sinx = 0$.

Multiplying the first equation by r_1, multiplying the second equation by r_2, and then subtracting the second equation from the first equation, we obtain $(r_1^2 + r_2^2) sinx = 0\ \forall x \in \mathbb{R}$ and for some $r_1, r_2 \in \mathbb{R}$. But then $r_1^2 + r_2^2 = 0$, which yields $r_1 = r_2 = 0$. Hence $\{sinx, cosx\}$ constitutes a basis of V.So $Dim_F V = 2$.

Definition(1.3.7). Suppose that $\{v_1, \cdots, v_\ell\}$ is a linearly independent set of a vector space V over F. If $\sum_{i=1}^{\ell} f_i v_i = \sum_{i=1}^{\ell} f_i' v_i$ for some $f_i, f_i' \in F$, we then have $\sum_{i=1}^{\ell}(f_i - f_i')v_i = \mathbb{O}$. It follows that $f_i - f_i' = 0 \ \forall \ i$, i.e., $f_i = f_i'$. So if $\{v_1, \cdots, v_\ell\}$ is a basis of V and $v = \sum_{i=1}^{\ell} f_i v_i$ for some $f_i's \in F$, then $f_i's$ are determined uniquely with respect to the basis. The uniquely determined ℓ-tuple $(f_1, \cdots, f_\ell)$ is called the *coordinate* of v with respect to the basis $\{v_1, \cdots, v_\ell\}$.

Each f_i is called the *i-th coordinate* of v or *i-th coordinate number* of v. Sporadically $f_i's$ are also called *coordinates* of v.

Example(1.3.8). Suppose that V is a vector space as in example1.3.6; V bears a resemblance to $\mathbb{R}^2$ because any vector $v \in V$ has the form $r_1 sinx + r_2 cosx$, so that (r_1, r_2) is just the coordinate of v with respect to the basis $\{sinx, cosx\}$.

Incidentally $\mathbb{R}^2$ may have a different basis from the ordinary natural basis consisting of standard unit vectors of $\mathbb{R}^2$.

Proposition(1.3.9). *Suppose that $\{g_1, \cdots, g_\ell\}$ generates a vector space V, i.e.,$\{g_1, \cdots, g_\ell\}$ is a set of generators of V.*

If $\{g_1, \cdots, g_i\} \subset \{g_1, \cdots, g_\ell\}$ is a maximal subset of linearly independent vectors, then V has $\{g_1, \cdots, g_i\}$ as a basis.

Proof. Our proof depends on showing that $\{g_1, \cdots, g_i\}$ generates V. So let V_G be the subspace (of V) generated by $\{g_1, \cdots, g_i\} =:$ G. If $V_G = V$, we are finished.

Otherwise $\forall v (\neq 0) \in V - V_G$, $f_1 g_1 + \cdots + f_i g_i + fv = 0$ for some $f_i's$ and f in F, so that we should have $f \neq 0$ because

$f = 0$ implies $\{g_1, \cdots, g_i, v\} \supsetneqq G$ is a maximal subset of linearly independent vectors which is a contradiction. But then $v = -\frac{f_1}{f} g_1 - \cdots - \frac{f_i}{f} g_i$ meaning that V is wholly generated by G. Hence our assertion holds.

Proposition(1.3.10). *Suppose that $B = \{b_1, \cdots, b_\ell\}$ is a basis of a vector space of V over F. For $m > \ell$, let $v_1, \cdots, v_\ell, \cdots, v_m$ be any vectors of V. Then $\{v_1, \cdots, v_\ell, \cdots, v_m\}$ becomes a linearly dependent set.*

Proof. Suppose that $\{v_1, \cdots, v_\ell, \cdots, v_m\}$ is a linearly independent set of V. If $v_1 = \sum_{i=1}^{\ell} f_{1i} b_i$ for some $f_{1i}'s \in F$, then some $f_{1i} \neq 0$. Without loss of generality we may assume $f_{11} \neq 0$. Then $b_1 = \frac{1}{f_{11}} v_1 - \frac{1}{f_{11}} \sum_{i=2}^{\ell} f_{1i} b_i$ implying $\{v_1, b_2, \cdots, b_\ell\}$ becomes a basis of V.

Proceeding likewise step by step, we may replace $b_2, \cdots, b_\ell$ by $v_2, \cdots, v_\ell$ without loss of generality until we exhaust all $b_i's$ and we make $\{v_1, \cdots, v_\ell\}$ become a basis of V. But then clearly $\{v_1, \cdots, v_\ell, \cdots, v_m\}$ is a linearly dependent set in V, a contradiction.

So we may see that dimension is well defined for finite dimension (see [8] in Exercises of this section). For infinite dimension, see theorem 2.6 chapter 4 [Hu].

Exercises of § 1-3

[1] Let V be a vector space of all continuous functions from $\mathbb{R}$ into $\mathbb{R}$. Show that e^x and e^{4x} are linearly independent vectors in the continuous function space V.

[2] Let $S = \{e^x, e^{4x}\}$ and V =the continuous function space as in [1]. What is V_S?

[3] Show that $\{e^x, 2e^x, e^{4x}\}$ cannot be a basis of V_S of [2].

[4] Show that $\{(1,1),(0,2)\}$ becomes a basis of $\mathbb{R}^2$.

[5] Find out the coordinates of $(2,4)$ with respect to bases $\{(1,0),(0,1)\}$, $\{(1,1),(0,2)\}$ respectively.

[6] What is $Dim_{\mathbb{R}}V$ for V in [1]?

[7] What is $Dim_{\mathbb{R}}V_S$ for $S = \{e^x, 2e^x, e^{4x}\}$ for V as in [1] ?

[8] Show that $Dim_F V$ is well defined in detail.

§ 1-4 Vector spaces operations

Useful operations among vector spaces appear ubiquitously in modern mathematics.

Example(1.4.1). Suppose that S and S' are subspaces of a vector space V over F. $S \cap S'$ is then a subspace of V evidently. For,$\forall f \in F$, $\forall x, y \in S \cap S'$, $fx \in S \cap S'$ and $x + y = S \cap S'$.

We shall call $S \cap S'$ the intersection space of S and S'. Inductively the intersection space of $S_1, \cdots, S_\ell$ is naturally defined.

Definition (1.4.2). For subspaces $S, S' \subset V$, we define $S + S' := \{x + x' \mid \forall x \in S, \ \forall x' \in S'\}$, which is called the *sum* of S and S'. Such an operation $+$ makes $S + S'$ a subspace of V. Inductively the sum of $S_1, \cdots, S_\ell$ of V is naturally defined.

Definition(1.4.3). For subspaces $S, S' \subset V$, if $S + S' = V$ and if any vector v of V is expressed uniquely as $v = x + x'$ for some $x \in S$ and $x' \in S$, then V is said to be the *direct sum* of S and S', denoted by $S \oplus S'$. Inductively the direct sum of subspaces $S_1, \cdots, S_\ell$ is easily defined.

The following proposition tells us when a sum of vector spaces is rendered a direct sum of them. Each component S_i of $S_1 \oplus \cdots \oplus S_\ell$ is often called a *direct summand* of it.

Proposition(1.4.4). *V becomes a direct sum of subspaces S and S' if and only if $S + S' = V$ and $S \cap S' = \{\mathbb{O}\}$.*

Proof. . (Only if part) is obvious by definition of direct sum.

(If part) Suppose that an arbitrary element $v \in V$ is expressed as $v = s_1 + s_1' = s_2 + s_2'$ for some $s_1, s_2 \in S$ and for some $s_1', s_2' \in S'$, so that $s_1 - s_2 = s_2' - s_1' \in S \cap S' = \{\mathbb{O}\}$. It follows that $s_1 = s_2$ and $s_1' = s_2'$, which ensures $V = S \oplus S'$.

Definition(1.4.5). Suppose that S and S' are any vector spaces over F. We define vector addition and multiplication by scalar on $S \times S'$ by the recipe:

$\forall (x, x') \in S \times S'$, $\forall (y, y') \in S \times S'$, $\forall f \in F$,

$$\begin{cases} (x, x') + (y, y') & := (x + y, x' + y'), \\ f(x, x') & := (fx, fx'). \end{cases}$$

In this case $S \times S'$ is rendered a vector space over F, which is called the *direct product* of S and S'. Very naturally the direct product of vector spaces $S_1, \cdots, S_\ell$ is defined in general.

Proposition(1.4.6).

(i) *Let $S_1, \cdots, S_\ell$ are subspaces of a vector space V; then we have $Dim_F(S_1 \oplus S_2 \oplus \cdots \oplus S_\ell) = Dim_F S_1 + \cdots + Dim_F S_\ell$.*

(ii) *For any vector spaces $S_1, \cdots, S_\ell$ over F, we have the formula $Dim_F(S_1 \times \cdots \times S_\ell) = Dim_F S_1 + \cdots + Dim_F S_\ell$.*

Proof. (i) If any of $S_1, \cdots, S_\ell$ is infinite dimensional, the formula follows from theorem 2.6 in chapter 4 [Hu] and from elementary set theory. If all $S_i's$ are finite dimensional, we let $\{b_{i1}, \cdots, b_{i\ell_i}\}$ be a basis of S_i $\forall i \in \{1, \cdots, \ell\}$. But then we can easily see $\bigcup_{i=1}^{\ell} \{b_{i1}, \cdots, b_{i\ell_i}\}$ becomes a basis of $\bigoplus_{i=1}^{\ell} S_i := S_1 \oplus \cdots \oplus S_\ell$. So (i) is clear.

(ii) By virtue of the above proof (i), it is also crystal clear that our formula holds.

Exercises of § 1-4

[1] Let $S_1, \cdots, S_\ell$ be subspaces of a vector space V over F. Show that $\exists$ bijection between $S_1 \times \cdots \times S_\ell$ and $S_1 \oplus \cdots \oplus S_\ell$.

[2] Verify that $\mathbb{R}^2 = \mathbb{R}(1,-1) \oplus \mathbb{R}(1,1)$.

[3] What is wrong with $\mathbb{Q}(1,-1) \oplus \mathbb{Q}(-1,1)$?

[4] Prove $\mathbb{C} = \mathbb{R} \oplus \mathbb{R}i$, where $i^2 = -1$.

[5] For any subspaces S_i for $i \in I(\neq \emptyset)$ of V, show that $\bigcap_{i\in I} S_i$ also becomes a subspace of V.

[6] Let S be a subspace of a vector space V over F. Define an equivalence relation $\sim$ on V as follows : $v \sim w$ if and only if $v - w \in S$.

Show that the set of equivalence classes forms a vector space over F. Such a vector space is called a *quotient* or *factor space* of V by S, denoted by V/S.

CHAPTER 2. OPERATIONS OF MATRICES

In order to study linear maps or simultaneous equations efficiently later, matrices are very useful for this purpose.

§ 2-1. Matrices Space

What is a matrix? We should define it clearly.

Definition(2.1.1). (i) By a *matrix* M over a field F, we mean an m by n (denoted $m \times n$) array of numbers in F as follows:

$$\begin{array}{ccc} & \begin{array}{ccccc} \phantom{f_{11}} & & j\ \text{th} & & \phantom{f_{1n}} \end{array} & \\ M = & \begin{pmatrix} f_{11} & \cdots & \vdots & \cdots & f_{1n} \\ \vdots & & \vdots & & \vdots \\ \cdot & \cdots & f_{ij} & \cdots & \cdots \\ \vdots & & \vdots & & \vdots \\ f_{m1} & \cdots & \vdots & \cdots & f_{mn} \end{pmatrix}_{m\times n} & \quad i\ \text{th} \end{array}$$

In this case we say the matrix M has m rows $M_1, \cdots, M_m \in F^n$ and n columns $M^1, \cdots, M^n$ or simply M is an m by n matrix. $M_i's$ are also called *row vectors* and $M^{i'}s$ *column vectors*. The

f_{ij} is called the (i, j)th *component* or (i, j)th *entry* of M, and usually f_{ij}(or f_{jk} etc.) denotes a general entry of the matrix M.

If $m = 1$, we also say that M is a *row vector*; if $n = 1$, we also say that M is a *column vector* in particular. If $m = n$, we may also say that M is an $m \times m$ *square matrix*. If all components are zero, we say M is a zero matrix written by $\mathbb{O}$.

(ii) For $M = (f_{ij})_{m\times n}$, the matrix ${}_tM := (f'_{ij})_{n\times m}$ with $f'_{ij} = f_{ji}$ is called the *transpose* of M. A square matrix M is called *symmetric* in case $M = {}_tM$ (see definition 2.1.3 (i) below).

(iii) If M is a square matrix like $M = (f_{ij})_{m\times m}$, we call $f'_{ii}s$ for $1 \leq i \leq m$ *diagonal entries* of M. In case all f_{ij} for $i \neq j$ are zero except possibly for $i = j$, we call M *a diagonal matrix*. We shall denote the diagonal matrix

$$\begin{pmatrix} f_{11} & & \bigcirc \\ & \ddots & \\ \bigcirc & & f_{mm} \end{pmatrix}$$

simply by diag $(f_{11} \cdots f_{mm})$.

In particular if a diagonal matrix has the diagonal line consisting of all $1's$, we call it an $m\times m$ *unit matrix* or $m\times m$ *identity matrix* denoted by I_m, i.e.,

$$I_m = \begin{pmatrix} 1 & 0 & \cdots & 0 \\ 0 & \ddots & & \vdots \\ \vdots & & \ddots & 0 \\ 0 & \cdots & 0 & 1 \end{pmatrix}_{m\times m} = diag(1\ 1 \cdots 1).$$

Example(2.1.2).

$$
{}_t\begin{pmatrix} 1 & 2 & 3 & 4 & 5 \\ 5 & 4 & 3 & 2 & 1 \end{pmatrix}_{2\times 5} = \begin{pmatrix} 1 & 5 \\ 2 & 4 \\ 3 & 3 \\ 4 & 2 \\ 5 & 1 \end{pmatrix}_{5\times 2} \quad ;
$$

$$
\begin{pmatrix} 1 & 2 & 3 \\ 4 & 5 & 6 \\ 7 & 8 & 9 \end{pmatrix}_{3\times 3} \qquad \text{is a } 3 \times 3 \text{ square matrix.}
$$

Definition(2.1.3). We may easily see that the set of all $m \times n$ matrices over F constitutes a vector space over F if we give the following prescriptions :

(i) equality : For $M = (f_{ij})_{m\times n}$ and $M' = (f'_{ij})_{m\times n}$, $M = M'$ if and only if $f_{ij} = f'_{ij}$ $\forall i$, $\forall j$;

(ii) vector addition : For $M = (f_{ij})_{m\times n}$ and $M' = (f'_{ij})_{m\times n}$, $M + M' := (f_{ij} + f'_{ij})_{m\times n}$;

(iii) multiplication by scalar : $\forall f \in F$, $\forall M = (f_{ij})_{m\times n}$, we define $f \cdot M := (f \cdot f_{ij})_{m\times n}$.

Proposition(2.1.4). *Elementary properties of transpose are at hand : $\forall f \in F$, $\forall m \times n$ matrices M, M', as above, we have*

(i) $M = {}_tM$ if and only if $f_{ij} = f_{ji}$ and $m = n$.

(ii) ${}_t(fM) = f \cdot {}_tM$;

(iii) ${}_t(M + M') = {}_tM + {}_tM'$.

Proof. Straightforward, so left to the reader.

Exercises of § 2-1

[1] If $M = (f_{ij})_{m \times n}$ and $M' = (f'_{jk})_{s \times t}$, where $f_{ij}, f'_{jk} \in F$, what is $M + M'$?

[2] With notations as above, what are ${}_tM$ and ${}_tM'$?

[3] What's the difference between two matrices $(f_{ij})_{m \times n}$ and $(f_{jk})_{m \times n}$?

[4] Show that the set of all $m \times n$ matrices over F constitutes a vector space of dimension $m \times n$.

[5] Let ${}_t(F^n)$ denote the set of all column vectors consisting of n-components in F. Show that ${}_t(F^n)$ forms a vector space.

[6] $(f_{ij})_{n \times n}$ with $f_{ij} = 0$ if $i \gneqq j$ is called a triangular matrix. Show that the set of all $n \times n$ triangular matrices forms a vector space over F.

[7] What if $f_{ij} = 0$ for $i \lneqq j$ in [6] ?

[8] If $M = (f_{ij})_{m \times m}$ for $f_{ij} \in F$, we then define the trace of M by : $tr(M) = \sum_{i=1}^{m} f_{ii}$.

(i) For any square matrices M, N of the same size, show $tr(MN) = tr(NM)$.

(ii) For any invertible N, show that $tr(N^{-1}MN) = tr(M)$ using (i).

§ 2-2. Matrices multiplication

In case two matrices have certain rows and columns, they may be multiplied as noted in the following

Definition(2.2.1). Suppose that $M = (f_{ij})_{m\times n}$ and $M' = (f'_{jk})_{n\times \ell}$ over F are given ; we then define $M \cdot M' := (f_{ij})_{m\times n} \cdot (f'_{jk})_{n\times \ell} := (f''_{ik})_{m\times \ell}$, where $f''_{ik} = \sum_{j=1}^{n} f_{ij} \cdot f'_{jk}$.

Example(2.2.2). Let us given two matrices

$$M = \begin{pmatrix} -1 & 2 & 3 \\ 4 & 5 & -6 \end{pmatrix}_{2\times 3}, M' = \begin{pmatrix} 7 & 8 & 9 & 10 \\ 9 & 8 & 7 & 6 \\ 5 & 4 & 3 & 2 \end{pmatrix}_{3\times 4} ;$$

we then compute

$$MM' = \begin{pmatrix} 26 & 20 & 14 & 8 \\ 43 & 48 & 53 & 58 \end{pmatrix}_{2\times 4}$$

as defined above.

Example(2.2.3). Suppose that we are given a system of linear equations like

$$\begin{cases} f_{11}x_1 + f_{12}x_2 + f_{13}x_3 = a_1 \\ f_{21}x_1 + f_{22}x_2 + f_{23}x_3 = a_2 \end{cases}$$

in indeterminates $x_i'^s$ over F. Using matrices multiplication, we render it a convenient form such as

$$\begin{pmatrix} f_{11} & f_{12} & f_{13} \\ f_{21} & f_{22} & f_{23} \end{pmatrix}_{2\times 3} \begin{pmatrix} x_1 \\ x_2 \\ x_3 \end{pmatrix}_{3\times 1} = \begin{pmatrix} a_1 \\ a_2 \end{pmatrix}_{2\times 1}.$$

Remark(2.2.4). For M and M' in Definition 2.2.1, we may express their multiplication in a different form using usual scalar product :

$$M \cdot M' = \begin{pmatrix} M_1 \cdot (M')^1 & \cdots & M_1 \cdot (M')^\ell \\ \vdots & \cdots & \vdots \\ M_m \cdot (M')^1 & \cdots & M_m \cdot (M')^\ell \end{pmatrix}_{m \times \ell} .$$

Proposition(2.2.5). *Elementary properties of matrices multiplication :*

(i) Suppose that two matrices M', M'' may be added and two matrices M, M' may be multiplied; we then have $M(M'+M'') = MM' + MM''$.

Similarly we have $(M'+M'')M = M'M + M''M$ if each term is defined.

(ii) Suppose that M, M' may be multiplied and M', M'' may be multiplied; then we have $(MM')M'' = M(M'M'')$, which is the associativity of matrices multiplication.

(iii) Suppose that M, M' may be multiplied; then $\forall f \in F$, $(fM)M' = M(fM') = f(MM')$.

(iv) For M and M' in (iii), ${}_t(MM') = {}_t(M') \cdot {}_tM$.

Proof. (i) Let $M = (f_{ij})_{m \times n}$, $M' = (f'_{jk})_{n \times \ell}$, and

$M'' = (f''_{jk})_{n \times \ell}$; we then have $M' + M'' = (f'_{jk} + f''_{jk})_{n \times \ell}$ and so

$$\begin{aligned}
M(M' + M'') &= \Big(\sum_{j=1}^{n} f_{ij}(f'_{jk} + f''_{jk})\Big)_{m\times\ell} \\
&= \Big(\sum_{j=1}^{n} f_{ij}f'_{jk}\Big)_{m\times\ell} + \Big(\sum_{j=1}^{n} f_{ij}f''_{jk}\Big)_{m\times\ell} \\
&= MM' + MM''.
\end{aligned}$$

(ii) Let $M = (f_{ij})_{m\times n}$, $M' = (f'_{jk})_{n\times s}$, and $M'' = (f''_{kh})_{s\times t}$; then both sides of our claim may be computed and compared like the following:

$$\begin{aligned}
\text{L.H.S.} &= \Big(\sum_{j=1}^{n} f_{ij}f'_{jk}\Big)_{m\times s} \cdot (f''_{kh})_{s\times t} \\
&(= ((i,k) - \text{th entry of } MM')_{m\times s} \cdot (f''_{kh})_{s\times t}) \\
&= \Big(\sum_{k=1}^{s}(\sum_{j=1}^{n} f_{ij}f'_{jk})f''_{kh}\Big)_{m\times t} \\
&(= ((i,h) - \text{th entry of } (M\cdot M')\cdot M'')_{m\times t}) \\
&= \Big(\sum_{k=1}^{s}\sum_{j=1}^{n} f_{ij}f'_{jk}f''_{kh}\Big)_{m\times t}.
\end{aligned}$$

$$
\begin{aligned}
\text{R.H.S.} &= (f_{ij})_{m\times n}\Big(\sum_{k=1}^{s} f'_{jk}f''_{kh}\Big)_{n\times t} \\
&\quad (= (f_{ij})_{m\times n}((j,h)-\text{th entry of } M'M'')_{n\times t}) \\
&= \Big(\sum_{j=1}^{n} f_{ij}(\sum_{k=1}^{s} f'_{jk}f''_{kh})\Big)_{m\times t} \\
&= \Big(\sum_{j=1}^{n}(\sum_{k=1}^{s} f_{ij}f'_{jk}f''_{kh})\Big)_{m\times t} \\
&= \Big(\sum_{j=1}^{n}\sum_{k=1}^{s} f_{ij}f'_{jk}f''_{kh}\Big)_{m\times t} \\
&= \Big(\sum_{k=1}^{s}\sum_{j=1}^{n} f_{ij}f'_{jk}f''_{kh}\Big)_{m\times t}.
\end{aligned}
$$

So we have L.H.S. = R.H.S.

(iii) For $M = (f_{ij})_{m\times n}$, $M' = (f'_{jk})_{n\times \ell}$ and $f \in F$, we have

$$
\begin{aligned}
(fM)M' &= (ff_{ij})_{m\times n}\cdot(f'_{jk})_{n\times\ell} \\
&= \Big(\sum_{j=1}^{n} ff_{ij}\cdot f'_{jk}\Big)_{m\times\ell} = \Big(\sum_{j=1}^{n} f_{ij}\cdot ff'_{jk}\Big)_{m\times\ell} \\
&= M\cdot(fM') = \Big(f\sum_{j=1}^{n} f_{ij}\cdot f'_{jk}\Big)_{m\times\ell} = f(MM').
\end{aligned}
$$

(iv) Let $M = (f_{ij})_{m\times n}$ and $M' = (f'_{jk})_{n\times\ell}$; then

$$
{}_t(M\cdot M') = {}_t\Big(\sum_{j=1}^{n} f_{ij}f'_{jk}\Big)_{m\times\ell} = (f''_{ki})_{\ell\times m}
$$

with $f''_{ki} = \sum_{j=1}^{n} f_{ij} f'_{jk}$, while

$$
\begin{aligned}
{}_tM' \cdot {}_tM &= (\hat{f}'_{kj})_{\ell\times n} \cdot (\hat{f}_{ji})_{n\times m} = \Big(\sum_{j=1}^{n} \hat{f}'_{kj}\hat{f}_{ji}\Big)_{\ell\times m} \\
(\text{with } \hat{f}'_{kj} = f'_{jk}, \hat{f}_{ji} = f_{ij}) &= \Big(\sum_{j=1}^{n} f'_{jk} \cdot f_{ij}\Big)_{\ell\times m} \\
&= \Big(\sum_{j=1}^{n} f_{ij} f'_{jk}\Big)_{\ell\times m}
\end{aligned}
$$

Hence our assertion follows.

Definition(2.2.6). If M is an $m \times m$ square matrix and if there is another $m \times m$ square matrix M' such that $MM' = M'M = I_m$, then M is called *an invertible* or *nonsingular matrix* and M' is called *the inverse (matrix)* of M. If there is no such M', we call M *a singular matrix.*

Proposition(2.2.7). *For M as the above definition 2.2.6, suppose that we have two $m \times m$ square matrices M', M'' such that $MM' = M''M = I_m$; we then have $M' = M''$, so that the inverse of M is uniquely determined.*

Proof. $M' = I_mM' = (M''M)M' = M''(MM')$ (by proposition 2.2.5)$= M''I_m = M''$.

Definition(2.2.8). Even if two matrices M, M' may be multiplied, both MM' and $M'M$ may not be computed in general. But if both of them may be computed and $MM' = M'M$, we then say M and M' *commute* (or are *commutative* or are *abelian*) with respect to matrix multiplication.

Exercises of § 2-2

[1] Compute $(MM')M''$ and $M(M'M'')$ in the following cases:

(i) $M = \begin{pmatrix} 2 & 1 & 3 \\ -1 & 1 & 2 \end{pmatrix}, M' = {}_t\begin{pmatrix} -1 & 0 & 1 \\ 2 & 3 & -1 \end{pmatrix},$
$M'' = {}_t(2 \;\;,\;\; 5).$

(ii) $M = {}_t\begin{pmatrix} 3 & 4 \\ 1 & 2 \end{pmatrix}, M' = {}_t\begin{pmatrix} 1 & 0 \\ -1 & 1 \end{pmatrix},$
$M'' = {}_t\begin{pmatrix} -2 & 3 \\ 4 & 5 \end{pmatrix}.$

[2] If $M = \begin{pmatrix} 2 & 1 & 0 \\ 1 & 0 & 0 \\ 0 & 0 & 0 \end{pmatrix}$ and $M' = \begin{pmatrix} 0 & -3 & 2 \\ 2 & 0 & -1 \\ 4 & 1 & 1 \end{pmatrix},$ compute $(M \pm M')^2$, MM', and $M'M$.

[3] Let E_{ij} be an $m \times m$ square matrix with (i, j)th entry 1 and all others zero.

Show that $E_{ij} \cdot E_{jk} = E_{ik}$ and $E_{ij} \cdot E_{hk} = 0$.

[4] Let $M = (f_{ij})_{m \times n}$ and $M' = (f'_{jk})_{n \times \ell}$ be given.

Express ${}_t(M \cdot M')$ using row vectors and column vectors of M and M'.

[5] Suppose that M_1 and M_2 are square matrices.

Verify that $M_1M_2 = M_2M_1$ if and only if $(M_1 + M_2)(M_1 - M_2) = M_1^2 - M_2^2$.

[6] Suppose that M_1, M_2, M_3 are matrices such that M_1M_2 and M_2M_3 are well defined.

Prove then that ${}_t(M_1M_2M_3) = {}_tM_3 \cdot {}_tM_2 \cdot {}_tM_1$.

[7] For $M_1 = \begin{pmatrix} 1 & 1 & 1 \\ 0 & 1 & 1 \\ 0 & 0 & 1 \end{pmatrix}$ and $M_2 = \begin{pmatrix} 0 & 1 & 1 \\ 0 & 0 & 1 \\ 0 & 0 & 0 \end{pmatrix}$, compute M_i^2 and M_i^3 respectively.

[8] Express scalar product on F^ℓ as matrices multiplication.

§ 2-3. System of linear equations

Suppose that we have a system of linear equations of the form

$$\left\{\begin{array}{c} f_{11}x_1 + \cdots + f_{1n}x_n = g_1 \\ f_{21}x_1 + \cdots + f_{2n}x_n = g_2 \\ \vdots \\ f_{m1}x_1 + \cdots + f_{mn}x_n = g_m \end{array}\right\} \qquad (*)$$

with n unknowns $x_i's$, where $f_{ij} \in F$.

Definition(2.3.1). If $(g_1, \cdots, g_m) = 0$, we call $(*)$ a system of *homogeneous linear equations.* Otherwise, we call $(*)$ a system of *inhomogeneous linear equations.*

Proposition(2.3.2). *The solution space S of $(*)$ when $g_1 = \cdots = g_m = 0$ may be interpreted in other ways like:*

(i) If $M := (f_{ij})_{m\times n}$ and $\mathbb{X} = {}_t(x_1, \cdots, x_n)$, then S consists of the solutions of $M \cdot \mathbb{X} = 0$.

(ii) S consists of all vectors orthogonal to row vectors $M_1, \cdots,$ M_m of M.

(iii) S consists of all solutions of the linear relations $\sum_{i=1}^{n} x_i M^i = 0$ between all columns of M.

Proof. Clear from definitions

Definition(2.3.3). For a given matrix $M = (f_{ij})_{m\times n}$, the dimension of the space generated by the columns $M^1, \cdots, M^n$ is called *the column rank* of M, while the dimension of the space generated by the rows $M_1, \cdots, M_m$ is called *the row rank* of M.

The reader will see soon that the column rank equals the row rank because the rank is closely related to the dimension of the solution space S.

Example(2.3.4). We want to find out the set of solutions of

$$\left\{ \begin{array}{c} x + 2y + z = 1 \\ x - z = 0 \end{array} \right\},$$

namely $\begin{pmatrix} 1 & 2 & 1 \\ 1 & 0 & -1 \end{pmatrix} {}_t(x, \ y, \ z) = {}_t(1, \ 0)$. If we put $z = x$ into the first equation, $2(x + y) = 1$ is obtained. Hence $\{(f, \frac{1}{2} - f, f) \mid \forall f \in F\}$ is obtained as the solution set. However this solution set may be interpreted as the set of all elements of the form $(0, \frac{1}{2}, 0) + f(1, -1, 1)$, $\forall f \in F$. We should notice here that $(0, \frac{1}{2}, 0)$ is just a solution of the given system of inhomogeneous linear equations, while $\{f(1, -1, 1) \mid \forall f \in F\}$ is just the solution space of the system of homogeneous linear equation $\begin{pmatrix} 1 & 2 & 1 \\ 1 & 0 & -1 \end{pmatrix} {}_t(x, \ y, \ z) = {}_t(0, \ 0)$. Obviously the solution space has 1 as its dimension, which is nothing else but (the number of unknowns$-$the rank of the matrix). See corollary 4.1.12 for detail.

The following proposition generalizes such a fact to a more general system of inhomogeneous linear equations.

Proposition(2.3.5). *Let* $M\mathbb{X} = G$ *be a system of inhomogeneous linear equations, where* $M = (f_{ij})_{m\times n}$, $\mathbb{X} = {}_t(x_1, \cdots, x_n)$, *and* $G = {}_t(g_1, \cdots, g_m) \in {}_t(F_m)$ *(refer to exercise 5 § 2-1 for notation).*

Suppose that $\mathbb{X}_0$ *is a particular solution of the above system and* $\widehat{\mathbb{X}}$ *is any solution of the system of homogeneous linear equations* $M\mathbb{X} = 0$. *Then any solution of* $M\mathbb{X} = G$ *is of the form* $\widehat{\mathbb{X}} + \mathbb{X}_0$ *and conversely* $\widehat{\mathbb{X}} + \mathbb{X}_0$ *is a general solution of* $M\mathbb{X} = G$,

Proof. Since $M\mathbb{X} = G$ and $M\mathbb{X}_0 = G$, we have $M(\mathbb{X} - \mathbb{X}_0) = G - G = 0$. Hence we may write $\mathbb{X} - \mathbb{X}_0 = \widehat{\mathbb{X}}$, i.e.,$\mathbb{X} = \widehat{\mathbb{X}} + \mathbb{X}_0$. Evidently the converse holds.

Exercises of § 2-3

[1] Compute the row ranks and the column ranks of the matrices $\begin{pmatrix} f_{11} & & \bigcirc \\ \vdots & \ddots & \\ f_{m1} & \cdots & f_{mm} \end{pmatrix} \begin{pmatrix} f_{11} & \cdots & f_{1m} \\ & \ddots & \vdots \\ \bigcirc & & f_{mm} \end{pmatrix}$, where $f_{ij} \in F$.

[2] Compute and compare row ranks and column ranks of the following matrices :

$$\text{(i)} \begin{pmatrix} 7 & 2 & -1 \\ 4 & 1 & 2 \end{pmatrix}, \text{(ii)} \begin{pmatrix} 7 & 3 \\ 3 & 4 \\ -4 & 3 \end{pmatrix}, \text{(iii)} \begin{pmatrix} -7 & 8 & 3 \\ 1 & 2 & 5 \\ 0 & 0 & 4 \end{pmatrix}.$$

[3] What is the dimension of the solution space of the following each system ?

$$\text{(i)} \begin{pmatrix} 7 & 2 & -1 \\ 4 & 1 & 2 \end{pmatrix} {}_t(x_1, \ x_2, \ x_3) = \mathbb{O},$$

$$\text{(ii)} \begin{pmatrix} 7 & 3 \\ 3 & 4 \\ -4 & 3 \end{pmatrix} {}_t(x_1, \ x_2) = \mathbb{O},$$

$$\text{(iii)} \begin{pmatrix} -7 & 8 & 3 \\ 1 & 2 & 5 \\ 0 & 0 & 4 \end{pmatrix} {}_t(x_1, \ x_2, \ x_3,) = \mathbb{O}.$$

[4] Find out the solution set of the following each system :

(i) $\begin{pmatrix} 7 & 2 & -1 \\ 4 & 1 & 2 \end{pmatrix} {}_t(x_1, \ x_2, \ x_3) = {}_t(2, \ 1),$

(ii) $\begin{pmatrix} 7 & 3 \\ 3 & 4 \\ -4 & 3 \end{pmatrix} {}_t(x_1, \ x_2) = {}_t(2, \ 1, \ 0),$

(iii) $\begin{pmatrix} -7 & 8 & 3 \\ 1 & 2 & 5 \\ 0 & 0 & 4 \end{pmatrix} {}_t(x_1, \ x_2, \ x_3) = {}_t(2, \ 0, \ -1).$

[5] Prove proposition 2.3.2 on your own and then discuss something about the solution with your friends.

CHAPTER 3. DETERMINANTS

If determinants were not invented, linear algebra would be in a hassle. In this chapter we introduce the concept of determinant and search for its important properties.

§ 3-1. Properties of determinants

Let us consider the following system of inhomogeneous linear equations

$$\begin{cases} f_{11}x_1 + f_{12}x_2 = g_1 \\ f_{21}x_1 + f_{22}x_2 = g_2 \end{cases},$$

where $f_{ij} \in F$ and $g_i \in F$.

Suppose $f_{11}f_{22} - f_{12}f_{21} \neq 0$; we then have the solution $x_1 = \dfrac{g_1f_{22} - g_2f_{12}}{f_{11}f_{22} - f_{12}f_{21}}$ and $x_2 = \dfrac{f_{11}g_2 - f_{21}g_1}{f_{11}f_{22} - f_{12}f_{21}}$. If we write the denominator as $\begin{vmatrix} f_{11} & f_{12} \\ f_{21} & f_{22} \end{vmatrix}$ and the numerators as $\begin{vmatrix} g_1 & f_{12} \\ g_2 & f_{22} \end{vmatrix}$ and $\begin{vmatrix} f_{11} & g_1 \\ f_{21} & g_2 \end{vmatrix}$ respectively, then we have simple expressions as

$$x_1 = \frac{\begin{vmatrix} g_1 & f_{12} \\ g_2 & f_{22} \end{vmatrix}}{\begin{vmatrix} f_{11} & f_{12} \\ f_{21} & f_{22} \end{vmatrix}}, \qquad x_2 = \frac{\begin{vmatrix} f_{11} & g_1 \\ f_{21} & g_2 \end{vmatrix}}{\begin{vmatrix} f_{11} & f_{12} \\ f_{21} & f_{22} \end{vmatrix}}.$$

Typeset by $\mathcal{A}\mathcal{M}\mathcal{S}$-TEX

At this moment we believe that there may exist some algorithm using such notations for the solution of a general system of linear equations of such kind. It is believed that such an idea brings forth the so called determinant. For the strict definition of this we should undergo some preliminary steps.

Definition(3.1.1). By a *permutation* of a set of integers $\{1, 2, \cdots, \ell\}$, we mean a one to one mapping from the set $\{1, 2, \cdots, \ell\}$ into itself. If i and j among elements of $\{1, 2, \cdots, \ell\}$ are mapped to each other and if the other rest elements are fixed, then we call such a permutation a *transposition* denoted by (i, j). In general a permutation is denoted by $\begin{pmatrix} 1 & 2 & \cdots & \ell \\ i_1 & i_2 & \cdots & i_\ell \end{pmatrix}$ or simply by $(i_1, i_2, \cdots, i_\ell)$.

If we denote $\{1, 2, \cdots, \ell\}$ by I_ℓ and if $\bar{\pi}$ is a permutation of I_ℓ, we define $\bar{\pi}^{-1}$ via $\bar{\pi}^{-1}(i) = j$ satisfying $\bar{\pi}(j) = i \ \forall i \in I_\ell$. Evidently $\bar{\pi}^{-1}$ also becomes a permutation, which is called the *inverse* (permutation) of $\bar{\pi}$.

Remark(3.1.2). If we denote the identity mapping by id, then $\bar{\pi}^{-1} \circ \bar{\pi} = \bar{\pi} \circ \bar{\pi}^{-1} = id$ holds. Also $(i, j)^2 = (i, j) \circ (i, j) = id$ and $(i, j)^{-1} = (i, j) = (j, i)$.

Proposition(3.1.3). *We may express any permutation of I_ℓ as a product of some transpositions of I_ℓ.*

Proof. Suppose that $\bar{\pi}$ is any permutation of I_ℓ. Since it is out of question for $\ell = 1$, we shall prove our assertion by induction on ℓ.

Assume that our claim holds for $\ell - 1 \geqq 1$. Let $\bar{\pi}(\ell) = k$. If $\ell = k$, we are done by induction hypothesis. If $\ell \neq k$, we consider a transposition $\tau = (\ell, k)$. Now if we compose $\bar{\pi}$ with τ, we have $\tau \circ \bar{\pi}(\ell) = \ell$, so that ℓ is fixed.

So we may view $\tau \circ \bar{\pi}$ as a permutation of $I_{\ell-1}$. Hence $\tau \circ \bar{\pi}$ may by expressed as a product of transpositions by induction hypothesis like $\tau \circ \bar{\pi} = \tau_1 \circ \cdots \circ \tau_m$, which yields $\bar{\pi} = \tau^{-1}\tau_1 \cdots \tau_m$ finishing our proof.

Proposition(3.1.4). *The number of transpositions whose composite expresses a permutation $\bar{\pi}$ of I_ℓ may change, but either even or odd is an invariant of $\bar{\pi}$.*

Proof. We consider the following product of differences of ℓ-variables $x_1, x_2, \cdots, x_\ell$:

$$
\begin{array}{rcccc}
P_d = (x_1 - x_2) & (x_1 - x_3) & \cdots & (x_1 - x_\ell) \\
 & (x_2 - x_3) & \cdots & (x_2 - x_\ell) \\
 & & & \vdots \\
 & & \ddots & \vdots \\
 & & & (x_{\ell-1} - x_\ell)
\end{array}
$$

If we apply a transposition $(1,2)$ to P_d, we have $-P_d$ evidently. In general if we apply a transposition (i,j) to P_d, we have $-P_d$. So if a permutation $\bar{\pi}$ decomposes into a product of transpositions, then the number of transpositions may vary but either even or odd should be invariant considering the sign of P_d which $\bar{\pi}$ is applied to.

Definition(3.1.5). In case a permutation becomes a product of even number of transpositions (odd number respectively), we call it an *even permutation (odd respectively).*

Definition(3.1.6). We introduce an ε-function from all the permutations of I_ℓ into $\{+1, -1\}$ defined by $\varepsilon(\bar{\pi}) = 1$ if $\bar{\pi}$ is even, and $\varepsilon(\bar{\pi}) = -1$ if $\bar{\pi}$ is odd.

Proposition(3.1.7). *ε has the following properties:*

(i) If we denote the permutation $\bar{\pi} = \begin{pmatrix} 1 & 2 & \cdots & \ell \\ i_1 & i_2 & \cdots & i_\ell \end{pmatrix}$ *by* $\bar{\pi} = (i_1 \cdots i_\ell)$ *and if* $(j_1 \cdots j_\ell)$ *is obtained from* $\bar{\pi}$ *by a transposition, we have* $\varepsilon(i_1 \cdots i_\ell) = -\varepsilon(j_1 \cdots j_\ell)$.

(ii) $\varepsilon(\bar{\pi}^{-1}) = \varepsilon(\bar{\pi})$.

(iii) $\varepsilon(1\ i_2 \cdots i_\ell) = \varepsilon(i_2 \cdots i_\ell)$.

Proof. (i) A transposition changes $\bar{\pi}$ an even permutation (resp. odd) to an odd permutation (resp. even). Hence (i) holds.

(ii) Because $\varepsilon(\bar{\pi} \circ \bar{\pi}^{-1}) = \varepsilon(\bar{\pi}) \circ \varepsilon(\bar{\pi}^{-1}) = \varepsilon(id) = 1$, $\varepsilon(\bar{\pi}) = \varepsilon(\bar{\pi}^{-1})$ follows.

(iii) $\varepsilon(1\ i_2 \cdots i_\ell) = \varepsilon((1\ \ i_2)(i_2\ \ 1)(i_2 \cdots i_\ell)) = \varepsilon(1\ \ i_2)\varepsilon(i_2\ \ 1)$ $\varepsilon(i_2 \cdots i_\ell) = \varepsilon(i_2 \cdots i_\ell)$.

Now we are ready to define the determinant of $\ell \times \ell$ matrix formally as follows.

Definition(3.1.8). Suppose that $M = (f_{ij})_{\ell \times \ell}$ with $f_{ij} \in F$ is an ℓ-square matrix; we then define its determinant, denoted by $|M|$ or $D(M)$, by $|M| = D(M) = \sum_{\bar{\pi}} \varepsilon(\bar{\pi}) f_{\bar{\pi}(1),1} \cdots f_{\bar{\pi}(\ell),\ell}$. We may also write $D(M) = D(M^1, \cdots, M^\ell)$ using columns of M.

Proposition(3.1.9). *D has the following properties:*

(i) $D : {}_tF^\ell \times \cdots \times {}_tF^\ell \longrightarrow F$ is a mapping such that if $M^j = aA + bB$ for $A, B \in {}_tF^\ell$ and $a, b \in F$, then $D(M) = aD(M^1, \cdots, A, \cdots, M^\ell) + bD(M^1, \cdots, B, \cdots, M^\ell)$, where A, B are situated in the j th column of M.

(ii) $D(I_\ell) = 1$ and $D(M^1, \cdots, M^{\ell-1}, \mathbb{O}) = 0$.

(iii) If two columns M^i, M^j are changed, then the determinant changes only its sign; e.g. $D(M^2, M^1, M^3, \cdots, M^\ell) = -D(M^1, M^2, M^3, \cdots, M^\ell) = -D(M)$. So if $M^i = M^j$, we have $D(M) = 0$ if characteristic $p \neq 2$. Also $D(M) = 0$ for $p = 2$.

(iv) $D(M^1, \cdots, M^i + fM^j, \cdots, M^\ell) = D(M)$ if i th column is added by fM^j for $i \neq j$ and $f \in F$.

(v) $D(M) = D({}_tM)$. So the above assertions also hold if columns are changed to rows.

Proof. (i) Suppose that ${}_tA = (a_{1j}, \cdots, a_{\ell j})$ and ${}_tB = (b_{1j}, \cdots, b_{\ell j})$; then by definition of determinant we have

$$
\begin{aligned}
&D(M^1, \cdots, aA + bB, \cdots, M^\ell) \\
=&\sum_{\overline{\pi}} \varepsilon(\overline{\pi}) f_{\overline{\pi}(1),1} \cdots f_{\overline{\pi}(j),j} \cdots f_{\overline{\pi}(\ell),\ell} \\
=&\sum_{\overline{\pi}} \varepsilon(\overline{\pi}) f_{\overline{\pi}(1),1} \cdots (a a_{\overline{\pi}(j),j} + b b_{\overline{\pi}(j),j}) \cdots f_{\overline{\pi}(\ell),\ell} \\
=&a\sum_{\overline{\pi}} \varepsilon(\overline{\pi}) f_{\overline{\pi}(1),1} \cdots a_{\overline{\pi}(j),j} \cdots f_{\overline{\pi}(\ell),\ell} \\
&+ b\sum_{\overline{\pi}} \varepsilon(\overline{\pi}) f_{\overline{\pi}(1),1} \cdots b_{\overline{\pi}(j),j} \cdots f_{\overline{\pi}(\ell),\ell} \\
=&aD(M^1, \cdots, A, \cdots, M^\ell) + bD(M^1, \cdots, B, \cdots, M^\ell).
\end{aligned}
$$

(ii)

$$\begin{aligned} D(I_\ell) &= \sum_{\bar{\pi}} \varepsilon(\bar{\pi}) f_{\bar{\pi}(1),1} \cdots f_{\bar{\pi}(j),j} \cdots f_{\bar{\pi}(\ell),\ell} \\ &= (\sum_{\bar{\pi}\neq id} 0) + \varepsilon(id) \cdot 1 = 1. \end{aligned}$$

Next we have

$$\begin{aligned} D(M^1, \cdots, M^{\ell-1}, 0) &= \sum_{\bar{\pi}} \varepsilon(\bar{\pi}) f_{\bar{\pi}(1),1} \cdots f_{\bar{\pi}(j),j} \cdots f_{\bar{\pi}(\ell),\ell} \\ &= \sum_{\bar{\pi}} \varepsilon(\bar{\pi}) \cdot 0 = 0. \end{aligned}$$

(iii) $D(M^2, M^1, \cdots, M^\ell) = \sum \varepsilon(i_1 i_2 \cdots i_\ell) f_{i_1,2} f_{i_2,1} f_{i_3,3} \cdots f_{i_\ell,\ell}$. Here if we put $i_1 = j_2, i_2 = j_1, i_3 = j_3, \cdots, i_\ell = j_\ell$, we obtain

$$\begin{aligned} &D(M^2, M^1, \cdots, M^\ell) \\ =&\sum \varepsilon(j_2 j_1 j_3 \cdots j_\ell) f_{j_2,2} f_{j_1,1} f_{j_3,3} \cdots f_{j_\ell,\ell} \\ =&-\sum \varepsilon(j_1 j_2 j_3 \cdots j_\ell) f_{j_1,1} \cdots f_{j_\ell,\ell} \\ =&-D(M^1, M^2, \cdots, M^\ell). \end{aligned}$$

Likewise we get the similar assertion in general if M^i and M^j are changed. So the latter assertion is also clear.

(iv) By (i) it is evident that

$$\begin{aligned} &D(M^1, \cdots, M^i + fM^j, \cdots, M^\ell) \\ =&D(M^1, \cdots, M^i, \cdots, M^\ell) \\ &+fD(M^1, \cdots, M^j, \cdots, M^j, \cdots, M^\ell) \\ =&D(M^1, \cdots, M^i, \cdots, M^\ell) = D(M). \end{aligned}$$

(v) By definition we have $D({}_tM) = \sum_{\bar\pi} \varepsilon(\bar\pi) f_{1,\bar\pi(1)} \cdots f_{\ell,\bar\pi(\ell)} = \sum_{\bar\pi} \varepsilon(\bar\pi^{-1}) f_{1,\bar\pi^{-1}(1)} \cdots f_{\ell,\bar\pi^{-1}(\ell)}$. Here if $\bar\pi^{-1}(j) = i$, then $\bar\pi(i) = j$.

Hence $D({}_tM) = \sum_{\bar\pi} \varepsilon(\bar\pi) f_{\bar\pi(1),1} \cdots f_{\bar\pi(\ell),\ell} = D(M)$.

Corollary(3.1.10). *Suppose that M, M' are ℓ-square matrices ; we then have $D(MM') = D(M) \cdot D(M')$.*

Poof. If we let $M = (f_{ij}), M' = (f'_{jk})$, then it is not difficult to see that

$$\begin{aligned} D(MM') &= D(f'_{11}M^1 + \cdots + f'_{\ell 1}M^\ell, \cdots, f'_{1\ell}M^1 + \cdots + f'_{\ell\ell}M^\ell) \\ &= \sum_{\bar\pi} f'_{\bar\pi(1),1} \cdots f'_{\bar\pi(\ell),\ell} D(M^{\bar\pi(1)}, \cdots, M^{\bar\pi(\ell)}) \\ &\qquad (\because \text{proposition 3.1.9 (i)}) \\ &= \sum_{\bar\pi} \varepsilon(\bar\pi) f'_{\bar\pi(1),1} \cdots f'_{\bar\pi(\ell),\ell} \cdot D(M) \\ &\qquad (\because \text{proposition 3.1.9 (iii)}) \\ &= D(M) \sum_{\bar\pi} \varepsilon(\bar\pi) f'_{\bar\pi(1),1} \cdots f'_{\bar\pi(\ell),\ell} = D(M) \cdot D(M'). \end{aligned}$$

Corollary(3.1.11). $D(M^{-1}) = D(M)^{-1}$ *if M is invertible.*

Proof. $1 = D(id) = D(M \cdot M^{-1}) = D(M) \cdot D(M^{-1})$, so that $D(M^{-1}) = D(M)^{-1}$ holds.

Now we sometimes need to rearrange all terms of $D(M)$ in terms of cofactors in what follows. First look at the definition

$$D(M) = D(f_{ij})_{\ell\times\ell} = \sum_{\bar\pi} \varepsilon(\bar\pi) f_{\bar\pi(1),1} \cdots f_{\bar\pi(\ell),\ell}$$

which may be viewed as a mapping of the jth column M^j with other columns fixed. Because each term in the summation has only one of the $f_{1j}, f_{2j}, \cdots, f_{\ell j}$ as a factor, we may collect all the terms with the common factor f_{ij} into one term like $f_{ij}M_{ij}$.

Hence it follows that $D(M) = \sum_{i=1}^{\ell} f_{ij}M_{ij}$, which may arise for any $j \in \{1, \cdots, \ell\}$. Here each M_{ij} is called *the cofactor* of f_{ij} with respect to $D(M)$.

Let m_{ij} denote the $(\ell - 1)$ square matrix obtained from M by deleting ith row and jth column. We then have M_{ij} explicitly as follows.

Proposition(3.1.12). *We have the relation* $M_{ij} = (-1)^{i+j} D(m_{ij})$ $\forall i, j \in \{1, \cdots, \ell\}$, *where* m_{ij} *is called* ij*th minor matrix and* $D(m_{ij})$ *is called* ij*th minor (determinant).*

Proof. We first consider M_{ij} for $i = j = 1$. Noting $D(M) = \sum_{\bar{\pi}} \varepsilon(\bar{\pi})\ f_{\bar{\pi}(1),1} \cdots f_{\bar{\pi}(\ell),\ell}$, we observe in this expression that a term with permutation $\bar{\pi}$ contains f_{11} as a factor exactly when $\bar{\pi}(1) = 1$.

So we have

$$\begin{aligned} M_{11} &= \sum_{\bar{\pi}(1)=1} \varepsilon(\bar{\pi}) f_{\bar{\pi}(2),2} \cdots f_{\bar{\pi}(\ell),\ell} \\ &= (-1)^{1+1} D(m_{11}) \\ &= D(m_{11}) \end{aligned}$$

because $\varepsilon(1, \bar{\pi}(2), \cdots, \bar{\pi}(\ell)) = \varepsilon(\bar{\pi}(2), \cdots, \bar{\pi}(\ell))$ (see proposition 3.1.7).

Next for any i, j, we consider the cofactor M_{ij}. We interchange column j with each preceding column successively until column j is first, and then similarly we do so with ith row until it is the first row.

Even if $D(m_{ij})$ is not changed by these actions, the sign of $D(M)$ does change by these. Exactly the sign changes a total of $i+j-2 = (i-1)+(j-1)$-times. But then it follows that $M_{ij} = (-1)^{i+j-2}D(m_{ij}) = (-1)^{i+j}D(m_{ij})$ because $M_{11} = D(m_{11})$.

Remark(3.1.13). From the preceding proposition we obtain $D(M) = \sum_{i=1}^{\ell}(-1)^{i+j}f_{ij}D(m_{ij})$, which is called the Laplace expansion of $D(M)$ according to the jth column.

Furthermore from proposition 3.1.9 (v), we obtain similarly $D(M) = \sum_{j=1}^{\ell}(-1)^{i+j}f_{ij}D(m_{ij})$, which is called the Laplace expansion of $D(M)$ according to the ith row.

Proposition(3.1.14). *With notations as above we have*

$$f_{1k}M_{1j} + f_{2k}M_{2j} + \cdots + f_{\ell k}M_{\ell j} = 0 \quad \text{if} \quad j \neq k;$$

we also have

$$f_{k1}M_{i1} + f_{k2}M_{i2} + \cdots + f_{k\ell}M_{i\ell} = 0 \quad \text{if} \quad k \neq i.$$

Proof. Because the first expansion is the (Laplace) expansion of the determinant of the matrix of the from

$$\begin{pmatrix} f_{11} & \cdots & f_{1k} & \cdots & f_{1k} & \cdots & f_{1\ell} \\ \vdots & & \vdots & & \vdots & & \vdots \\ f_{\ell 1} & \cdots & f_{\ell k} & \cdots & f_{\ell k} & \cdots & f_{\ell\ell} \end{pmatrix}$$

whose jth column and kth column are equal, it becomes zero by virtue of proposition 3.1.9 9 (iii).

Likewise the latter expression is just the (Laplace) expansion of the determinant of the matrix whose ith row and kth row are equal, so that it becomes 0 evidently.

Now we may compute determinants easily if we make use of many properties and facts of them up to now.

Example(3.1.15). Let us calculate the following determinant:

$$d := \begin{vmatrix} 1 & 2 & 1 & 4 \\ 3 & 1 & -1 & 1 \\ 1 & 5 & 2 & -3 \\ 1 & 2 & 3 & 7 \end{vmatrix}$$

First add the third column to the 2nd column; afterwards add the 3rd column to the 4th column. So we get

$$d = \begin{vmatrix} 1 & 3 & 1 & 4 \\ 3 & 0 & -1 & 1 \\ 1 & 7 & 2 & -3 \\ 1 & 5 & 3 & 7 \end{vmatrix} = \begin{vmatrix} 1 & 3 & 1 & 5 \\ 3 & 0 & -1 & 0 \\ 1 & 7 & 2 & -1 \\ 1 & 5 & 3 & 10 \end{vmatrix}.$$

Next the last one reduces to $\begin{vmatrix} 4 & 3 & 1 & 5 \\ 0 & 0 & -1 & 0 \\ 7 & 7 & 2 & -1 \\ 10 & 5 & 3 & 10 \end{vmatrix}$ three times the 3rd column being added to the 1st column. If we expand this according to the 2nd row, we obtain $\begin{vmatrix} 4 & 3 & 5 \\ 7 & 7 & -1 \\ 10 & 5 & 10 \end{vmatrix}$; if two times the 2nd column is subtracted from the 1st and 3rd column of the last determinant, we get $\begin{vmatrix} -2 & 3 & -1 \\ -7 & 7 & -15 \\ 0 & 5 & 0 \end{vmatrix}$ which reduces to $-5\begin{vmatrix} -2 & -1 \\ -7 & -15 \end{vmatrix} = -5 \times 23 = -115$ by Laplace expansion according to 3rd row.

Example(3.1.16). If we expand the determinant

$$d := \begin{vmatrix} 7 & 3 & 1 \\ 2 & -4 & 3 \\ 4 & 5 & -2 \end{vmatrix}$$

according to the first column, we have

$$\begin{aligned} d &= 7\begin{vmatrix} -4 & 3 \\ 5 & -2 \end{vmatrix} - 2\begin{vmatrix} 3 & 1 \\ 5 & -2 \end{vmatrix} + 4\begin{vmatrix} 3 & 1 \\ -4 & 3 \end{vmatrix} \\ &= 7(8-15) - 2(-6-5) + 4(9+4) \\ &= 25. \end{aligned}$$

Exercises of § 3-1

[1] For any numbers $x_1, \cdots, x_\ell$, put

$$V_\ell := \begin{vmatrix} 1 & 1 & \cdots & 1 \\ x_1 & x_2 & \cdots & x_\ell \\ \vdots & \vdots & & \vdots \\ x_1^{\ell-1} & x_2^{\ell-1} & \cdots & x_\ell^{\ell-1} \end{vmatrix}$$

which is called *the Vandermonde determinant.* Verify by induction that $V_\ell = (x_\ell - x_1)\cdots(x_2 - x_1)V_{\ell-1} = \prod_{j>i}(x_j - x_i)$ holds.

[2] Compute the determinants like $|\,xI_\ell\,|$ and $|diag(x_{11}, x_{22}, \cdots, x_{\ell\ell})|$, where $x, x'_{ii}s$ are numbers in F.

[3] Compute $D(M)$ for the following matrices:

$$\begin{pmatrix} f_{11} & f_{12} & \cdots & f_{1\ell} \\ 0 & f_{22} & \cdots & f_{2\ell} \\ \vdots & \vdots & \ddots & \vdots \\ 0 & 0 & \cdots\ 0 & f_{\ell\ell} \end{pmatrix},$$

$$\begin{pmatrix} f_{11} & 0 & \cdots & 0 \\ f_{21} & f_{22} & \cdots & 0 \\ \vdots & \vdots & \ddots & \vdots \\ \vdots & \vdots & & 0 \\ f_{m1} & f_{m2} & \cdots & f_{mm} \end{pmatrix},$$

$$\begin{pmatrix} cos\theta & sin\theta \\ -\sin\theta & cos\theta \end{pmatrix}.$$

[4] Show that $\begin{vmatrix} 1 & x & y & z+w \\ 1 & y & z & w+x \\ 1 & z & w & x+y \\ 1 & w & x & y+z \end{vmatrix} = 0,$

[5] Prove that

$$\frac{d}{dx}\begin{vmatrix} f_{11}(x) & f_{12}(x) & f_{13}(x) \\ f_{21}(x) & f_{22}(x) & f_{23}(x) \\ f_{31}(x) & f_{32}(x) & f_{33}(x) \end{vmatrix} = \begin{vmatrix} f'_{11}(x) & f_{12}(x) & f_{13}(x) \\ f'_{21}(x) & f_{22}(x) & f_{23}(x) \\ f'_{31}(x) & f_{32}(x) & f_{33}(x) \end{vmatrix}$$

$$+\begin{vmatrix} f_{11}(x) & f'_{12}(x) & f_{13}(x) \\ f_{21}(x) & f'_{22}(x) & f_{23}(x) \\ f_{31}(x) & f'_{32}(x) & f_{33}(x) \end{vmatrix} + \begin{vmatrix} f_{11}(x) & f_{12}(x) & f'_{13}(x) \\ f_{21}(x) & f_{22}(x) & f'_{23}(x) \\ f_{31}(x) & f_{32}(x) & f'_{33}(x) \end{vmatrix},$$

where $f_{ij}(x)'s$ are differentiable functions of variable x.

[6] Compute the following:

(i) $\varepsilon(2\ 6\ 7\ 3\ 1\ 4\ 5)$, $\varepsilon(2\ 3\ 5\ 4\ 1)$, $\varepsilon(5\ 3\ 4\ 1\ 2\ 6)$.

(ii) $\begin{vmatrix} 2 & 1 & 3 \\ -1 & 2 & 1 \\ 4 & 3 & 5 \end{vmatrix}$, $\begin{vmatrix} -1 & 0 & 3 & 0 \\ 1 & 3 & 1 & 4 \\ 2 & 2 & 5 & 1 \\ 0 & 1 & 7 & 2 \end{vmatrix}$, $\begin{vmatrix} 3 & 5 & 7 \\ 1 & 5 & 8 \\ 1 & 2 & 8 \end{vmatrix}$.

[7] Prove proposition(3.1.9) (iii) in detail.

§ 3-2. The rule of Cramer

As noted in §3-1, certain systems of linear equations may be solved by way of properties of determinants. So we offer the readers the so called Cramer's formula in the first place.

Theorem(3.2.1) (Cramer's formula). *Suppose that* $M = (f_{ij})_{\ell\times\ell}$ *is an* ℓ*-square matrix with* $D(M) \neq 0$ *and* N *is a column vector. Let* $\sum_{j=1}^{\ell} x_j M^j = N$ *holds for some numbers* $x_j \in F$ $\forall j \in \{1, \cdots, \ell\}$. *We then have* $x_j = \dfrac{D(M^1, \cdots, N, \cdots, M^\ell)}{D(M)}$, *where* N *is situated in the* j*th column of the numerator.*

Proof. Because we have a system of ℓ linear equations in ℓ unknowns like

$$\left\{\begin{matrix} x_1 f_{11} + \cdots + x_\ell f_{1\ell} = b_1 \\ x_1 f_{21} + \cdots + x_\ell f_{2\ell} = b_2 \\ \vdots \quad\quad \vdots \quad\quad \vdots \quad\quad \vdots \\ x_1 f_{\ell 1} + \cdots + x_\ell f_{\ell\ell} = b_\ell \end{matrix}\right\} \quad \text{if and only if} \quad \sum_{j=1}^{\ell} x_j M^j = N,$$

we may compute $D(M^1, \cdots, N, \cdots, M^\ell)$ for x_j as follows.

$$
\begin{aligned}
&D(M^1, \cdots, N, \cdots, M^\ell) \\
=&D(M^1, \cdots, \sum_{i=1}^{\ell} x_i M^i, \cdots, M^\ell) \\
=&D(M^1, \cdots, x_1 M^1, \cdots, M^\ell) \\
&+ \cdots + D(M^1, \cdots, x_j M^j, \cdots, M^\ell) \\
&+ \cdots + D(M^1, \cdots, x_\ell M^\ell, \cdots, M^\ell) \ (\because \text{proposition 3.1.9}) \\
=&x_1 D(M^1, \cdots, M^1, \cdots, M^\ell) \\
&+ \cdots + x_j D(M^1, \cdots, M^j, \cdots, M^\ell) \\
&+ \cdots + x_\ell D(M^1, \cdots, M^\ell, \cdots, M^\ell) \\
=&x_j D(M^1, \cdots, M^j, \cdots, M^\ell) \ (\because \text{proposition 3.1.9})
\end{aligned}
$$

$$\therefore x_j = \frac{D(M^1, \cdots, N, \cdots, M^\ell)}{D(M)} \text{ as desired.}$$

Proposition(3.2.2). *Suppose that $M^1, \cdots, M^\ell$ (resp.$M_1, \cdots, M_\ell$) are linearly dependent; we then have $D(M) = 0$. So if $D(M) \neq 0$, then $M^1, \cdots, M^\ell$ (resp.$M_1, \cdots, M_\ell$) are linearly independent.*

Proof. Let there be a linearly dependent relation $\sum_{j=1}^{\ell} f_j M^j = 0$. Without loss of generality we may assume $f_1 \neq 0$. So it follows that $M^1 = \sum_{j=2}^{\ell}(-\frac{f_j}{f_1})M^j$. Substituting the summation for M^1 in $D(M)$, we have $D(M) = D(M^1, \cdots, M^\ell) = D(\sum_{j=2}^{\ell}(-\frac{f_j}{f_1})M^j, \cdots, M^\ell) = 0$ by virtue of proposition 3.1.9 (iii).

Corollary(3.2.3). *Suppose that $D(M) \neq 0$ and N is any column vector in F^ℓ; then N is just a linear combination of $M^{i'}s$. (Resp.M'_is are also ok if N is any row vector).*

Proof. By the preceding proposition 3.2.2, we see $M^{j'}s$ are linearly independent in F^ℓ. Hence it follows that $M^1, \cdots, M^\ell$ are generators of F^ℓ, i.e., $\{M^1, \cdots, M^\ell\}$ forms a basis of F^ℓ.

Example(3.2.4). We want to use the Cramer's rule to solve the system of inhomogeneous linear equations of the form

$$7x_1 + 3x_2 + 4x_3 = -1$$

$$4x_1 + 2x_2 + 3x_3 = 0$$

$$-x_1 + x_2 + 2x_3 = 3$$

By this formula it follows that

$$x_1 = \frac{\begin{vmatrix} -1 & 3 & 4 \\ 0 & 2 & 3 \\ 3 & 1 & 2 \end{vmatrix}}{\begin{vmatrix} 7 & 3 & 4 \\ 4 & 2 & 3 \\ -1 & 1 & 2 \end{vmatrix}}, \quad x_2 = \frac{\begin{vmatrix} 7 & -1 & 4 \\ 4 & 0 & 3 \\ -1 & 3 & 2 \end{vmatrix}}{\begin{vmatrix} 7 & 3 & 4 \\ 4 & 2 & 3 \\ -1 & 1 & 2 \end{vmatrix}}, \quad x_3 = \frac{\begin{vmatrix} 7 & 3 & -1 \\ 4 & 2 & 0 \\ -1 & 1 & 3 \end{vmatrix}}{\begin{vmatrix} 7 & 3 & 4 \\ 4 & 2 & 3 \\ -1 & 1 & 2 \end{vmatrix}}.$$

Exercises of § 3-2

[1] Find out solutions of the following system of linear equations:

$$\text{(i)}\ \begin{cases} -3x - 2y + z = 1 \\ -x + y - z = 4 \\ -2x - 3y + z = 2 \end{cases} \qquad \text{(ii)}\ \begin{cases} x_1 + \alpha x_2 + \alpha^2 x_3 = \alpha^3 \\ x_1 + \beta x_2 + \beta^2 x_3 = \beta^3 \\ x_1 + \gamma x_2 + \gamma^2 x_3 = \gamma^3 , \end{cases}$$

where α, β, γ are distinct.

[2] What are solutions of the following system of inhomogeneous linear equations:

$$\text{(i)}\begin{cases} w + 2x - 3y + 4z = 1 \\ -2w + x - 4y - z = 0 \\ w + x + y + z = 0 \\ -w - x - y + z = 3 \end{cases} \qquad \text{(ii)}\begin{cases} -x_1 - x_2 - x_3 + x_4 = 4 \\ x_1 + x_2 + 2x_3 - 5x_4 = 0 \\ 2x_1 + x_2 + 3x_3 + 5x_4 = 1 \\ -2x_1 - x_2 + 4x_3 + x_4 = 9 \end{cases}$$

[3] Let the determinant of the coefficients of a system of linear equations be of the form $\begin{vmatrix} X & Y \\ O & Z \end{vmatrix}$, where X, Y and Z are ℓ-square matrices.

Show then that $\begin{vmatrix} X & Y \\ O & Z \end{vmatrix} = |XZ| = |X| \cdot |Z|$ holds.

§ 3-3. Invertible matrix

We should like to know how to obtain the inverse (matrix) of a given invertible matrix. First we introduce a definition.

Definition(3.3.1). Let M_{ij} be the cofactor of f_{ij} for a determinant $|(f_{ij})| = |M|$. Consider the matrix ${}_t(M_{ij})$ whose ijth component or entry is just M_{ji}. We call such a matrix ${}_t(M_{ij})$ *the adjoint matrix* of M and write it as *adj* M.

Proposition(3.3.2). *Let M be an invertible matrix; then M^{-1} $= adj\ M/D(M)$ holds. Furthermore M^{-1} is uniquely determined.*

Proof. Since $M \cdot M^{-1} = I_\ell$ for some $\ell \in \mathbb{N}$, it follows that $D(M \cdot M^{-1}) = D(M) \cdot D(M^{-1}) = D(I_\ell)$. So it yields $D(M) \cdot D(M^{-1}) = 1 \neq 0$ and hence $D(M) \neq 0$.

Next we consider $M \cdot adj\ M$, which results in a matrix of the form

$$\begin{pmatrix} D(M) & & \bigcirc \\ & \ddots & \\ \bigcirc & & D(M) \end{pmatrix} = diag(D(M), \cdots, D(M)) = D(M) \cdot I_\ell$$

by virtue of proposition 3.1.14. Similarly $adj\ M \cdot M = D(M) \cdot I_\ell$. Hence we obtain $M^{-1} = \frac{1}{D(M)} \cdot adj\ M$. Proposition 2.2.7 guarantees its uniqueness.

Remark(3.3.3). By virtue of proposition 3.1.12, and the preceding proposition 3.3.2, the following formula reassures us :

$M^{-1} = {}_t\Big((-1)^{i+j}D(m_{ij})/D(M)\Big)$. Furthermore it is evident that M is invertible if and only if $D(M) \neq 0$.

Example(3.3.4). Suppose that $|M| = \begin{vmatrix} \alpha & \beta \\ \gamma & \delta \end{vmatrix} \neq 0$; we then have $M_{11} = \delta, M_{12} = -\gamma, M_{21} = -\beta, M_{22} = \alpha$. So we obtain $adj\ M = \begin{pmatrix} \delta & -\beta \\ -\gamma & \alpha \end{pmatrix}$.

Hence $M^{-1} = {}_t(M_{ij}/D(M)) = \frac{1}{\alpha\delta-\beta\gamma} \cdot \begin{pmatrix} \delta & -\beta \\ -\gamma & \alpha \end{pmatrix}$ is yielded.

Example(3.3.5). We'd like to search for the inverse (matrix) of $M = \begin{pmatrix} \alpha_{11} & \alpha_{12} & \alpha_{13} \\ 0 & \alpha_{22} & \alpha_{23} \\ 0 & 0 & \alpha_{33} \end{pmatrix}$. Now

$$M_{11} = \alpha_{22}\alpha_{33}, M_{12} = 0, M_{13} = 0,$$

$$M_{21} = -\alpha_{12}\alpha_{33}, M_{22} = \alpha_{11}\alpha_{33}, M_{23} = 0,$$

$$M_{31} = \alpha_{12}\alpha_{23} - \alpha_{22}\alpha_{13}, M_{32} = -\alpha_{11}\alpha_{23}, M_{33} = \alpha_{11}\alpha_{22},$$

and $|M| = \alpha_{11}\alpha_{22}\alpha_{33}$. Hence we get

$$M^{-1} = \alpha_{11}^{-1}\alpha_{22}^{-1}\alpha_{33}^{-1} \begin{pmatrix} \alpha_{22}\alpha_{33} & -\alpha_{12}\alpha_{33} & \alpha_{12}\alpha_{23} - \alpha_{22}\alpha_{13} \\ 0 & \alpha_{11}\alpha_{33} & -\alpha_{11}\alpha_{23} \\ 0 & 0 & \alpha_{11}\alpha_{22} \end{pmatrix}.$$

Proposition(3.3.6). *Let there be a system of ℓ-homogeneous linear equations with ℓ-unknowns. Then this system has a non-trivial solution (,i.e. a solution which is dfferent from $\mathbb{O}$ solution called a trivial solution) if and only if the determinant of the matrix M consisting of all the coefficients of the given system should be zero.*

Proof. Only if part is just the contrapositive of Cramer's theorem 3.2.1. Next for the if part, we observe $D(M) = 0$ if and only if M is not invertible by virtue of remark 3.3.3. Suppose that $M^1, \cdots, M^\ell$ are linearly independent in ${}_t(F^\ell)$ which denotes the set of all column vectors consisting of ℓ-components in F. We shall use this notation freely henceforth when necessary.

Now if we write ${}_tU_i =: U^i$ for $i \in \{1, \cdots, \ell\}$, then each U^i becomes a linear combination of $M^1, \cdots, M^\ell$. So this means that we have a matrix M' such that $M(M')^j = U^j \ \forall j \in \{1, \cdots, \ell\}$. Hence it follows that $MM' = I_\ell$, so that $D(M) \neq 0$. From remark 3.3.3 we then obtain M being invertible.

So by contrapositive, we understand that M being not invertible implies $M^1, \cdots, M^\ell$ are linearly dependent. So after all we must have a nontrivial solution for the given system of homogeneous linear equations.

Exercises of § 3-3

[1] Find out the inverses, if any, of the following matrices:

$$\text{(i)} \begin{pmatrix} 1 & 1 & 2 \\ 1 & -3 & 0 \\ -2 & 1 & 4 \end{pmatrix} \quad \text{(ii)} \begin{pmatrix} 5 & 1 & 2 \\ -1 & -2 & 1 \\ 3 & -4 & -2 \end{pmatrix}$$

$$\text{(iii)} \begin{pmatrix} 1 & -2 & 0 \\ 0 & 3 & 2 \\ -3 & 4 & -1 \end{pmatrix}$$

[2] What are the inverses of the following 2×2 matrices ?

$$\text{(i)} \begin{pmatrix} -4 & 2 \\ 3 & -1 \end{pmatrix} \text{(ii)} \begin{pmatrix} 2 & 3 \\ 4 & -5 \end{pmatrix} \text{(iii)} \begin{pmatrix} 3 & -5 \\ 2 & 1 \end{pmatrix}$$

[3] Solve the system of ℓ-linear equations with ℓ-unknowns in the following form: $M\mathbb{X} = G$ as noted in proposition 2.3.5, where $D(M) \neq 0$.

[4] Show that $({}_tM)^{-1} = {}_t(M^{-1})$ holds.

[5] Solve the following systems using invertible matrices:

$$\text{(i)} \begin{cases} 2x + y = 0 \\ 4x + 3y = -1 \end{cases} \quad \text{(ii)} \begin{cases} 3x - y = 2 \\ 4x + y = 1 \end{cases}$$

$$\text{(iii)} \begin{cases} x + 2y = 3 \\ 4x + 5y = -6 \end{cases}$$

[6] Let M, M' be invertible $\ell \times \ell$ matrices; show then that MM' is also invertible and $(MM')^{-1} = (M')^{-1}M^{-1}$.

§ 3-4. Matrix rank

We have a close relationship of a solution space of a given system with the maximal number of linearly independent column vectors (or row vectors) of the coefficients matrix of the given system. We shall learn this fact soon in what follows.

Definition(3.4.1). Let us given an $m \times n$ matrix

$$M = \begin{pmatrix} f_{11} & \cdots & f_{1n} \\ \vdots & \vdots & \vdots \\ f_{m1} & \cdots & f_{mn} \end{pmatrix}$$

with entries in F. We remember that the dimension of the vector space generated by row rectors $M_1, \cdots, M_m$ is called *the row rank* of the matrix M, while the dimension of the vector space generated by column vectors $M^1, \cdots, M^n$ is called *the column rank* of M. We shall see, however, that these ranks are equal. So the rank of a matrix is well-defined. For its proof we consider the following elementary operations for now.

EC(i) :Exchange two column vectors.

EC(ii) :Multiply a column vector by any nonzero scalar.

EC(iii) :Multiply a column vector by any scalar and then add it to another column vector.

ER(i) :Exchange two row vectors.

ER(ii) :Multiply a row vector by any nonzero scalar.

ER(iii) :Multiply a row vector by any scalar and then add it to another row vector.

Proposition(3.4.2). *Let $M = (f_{ij})$ be any $m \times n$ matrix. The column rank and row rank of M does not change by any operation of EC(i) through ER(iii).*

Proof. **Step I**. ER(i) through ER(iii) does not change the row rank: It is evident by definition of row rank that ER(i) does not change the row rank. ER(ii) also does not change the row rank because all linear combinations of $M_1, \cdots, M_m$ equal all linear combinations of $f_1M_1, \cdots, f_mM_m$ for any nonzero $f_i's \in F$.

ER(iii) also does not change the row rank because

$$\begin{aligned} &f_1M_1 + \cdots + f_i(M_i + fM_j) + \cdots + f_jM_j + \cdots + f_mM_m \\ =&f_1M_1 + \cdots + f_iM_i + \cdots + (f_j + ff_i)M_j + \cdots + f_mM_m \end{aligned}$$

holds for $f, f_i's \in F$.

Step II. ER(i) through ER(iii) does not change the column rank : Suppose that there is a linear (in)dependence relation $f_1M^1 + \cdots + f_nM^n = 0$. But this means that we have a solution vector $(f_1, \cdots, f_\ell)$ satisfying a system of homogeneous linear equations of the form

$$\begin{cases} f_{11}x_1 + f_{12}x_2 + \cdots + f_{1n}x_n = 0 \\ \quad\vdots \qquad\quad \vdots \qquad\qquad\quad \vdots \\ f_{m1}x_1 + f_{m2}x_2 + \cdots + f_{mn}x_n = 0. \end{cases}$$

In this situation ER(i) just means two equations of this system are changed in order, so linear (in)dependence of $M^1, \cdots, M^n$ is exactly the same as that of $(M')^1, \cdots, (M')^n$, where M' is the matrix obtained by ER(i) acted on M.

Next let M'' be the matrix obtained by ER(ii) acted on M. Then we obtain a system of linear equations similarly as above:

$$\begin{cases} f_{11}x_1 + f_{12}x_2 + \cdots + f_{1n}x_n = 0 \\ \qquad \cdots\cdots \\ ff_{i1}x_1 + ff_{i2}x_2 + \cdots + ff_{in}x_n = 0 \\ \qquad \cdots\cdots \\ f_{m1}x_1 + f_{m2}x_2 + \cdots + f_{mn}x_n = 0, \end{cases}$$

where $f \neq 0$ and only ith equation is different from that of the former system. Also in this case the solution space is absolutely equal to the solution space of $M \cdot \mathbb{X} = \mathbb{O}$ which is the equation as in proposition 2.3.2(i). So ER(ii) does not change the column rank of M. Now consider ER(iii) acted on M. All equations are then the same except for $(f_{i1} + ff_{j1})x_1 + (f_{i2} + ff_{j2})x_2 + \cdots + (f_{in} + ff_{jn})x_n = 0$ when we think of the system of linear equations as above. But both systems of linear equations have the same solution space because

$$\begin{aligned} &(f_{i1} + ff_{j1})x_1 + (f_{i2} + ff_{j2})x_2 + \cdots + (f_{in} + ff_{jn})x_n \\ =& \sum_{k=1}^{n} f_{ik}x_k + f\sum_{k=1}^{n} f_{jk}x_k = 0 \end{aligned}$$

holds. Hence ER(iii) also does not change the column rank of M.

Step III. EC(i) through EC(iii) does not change the row rank and the column rank : It is evident that this fact is right considering ${}_tM$ instead of M and all arguments here are just the repetition of the former ones.

Proposition(3.4.3). *Any matrix $M = (f_{ij})_{m\times n}$ has the same column rank as the row rank.*

Proof. Since 1×1 case is trivial, we proceed by induction on $m \times n$. We have only to show that M reduces to the form $\hat{M} = (\delta_{ij})_{r\times r}$ with the rest entries zero by applying ER(i) through ER(iii) and EC(i) through EC(iii) to $M = (f_{ij})_{m\times n}$, where $\delta_{ij} = 1$ if $i = j$ and 0 otherwise. Without loss of generality we may assume that $f_{11} \neq 0$ because we may use both EC(i) and ER(i) until the first entry of the changed matrix with the same rank as M becomes nonzero. Next if we multiply the 1st row by some numbers and substract them from the rest rows, we may make M into the form $M' = \begin{pmatrix} f_{11} & f_{12} & \cdots & f_{1n} \\ 0 & g_{22} & \cdots & g_{2n} \\ \vdots & \vdots & & \vdots \\ 0 & g_{m2} & \cdots & g_{mn} \end{pmatrix}$. Next if we multiply the 1st column by some numbers and subtract them from the rest columns we may also reduce M' to the form $M'' = \begin{pmatrix} f_{11} & 0 & \cdots & 0 \\ 0 & g_{22} & \cdots & g_{2n} \\ \vdots & \vdots & & \vdots \\ 0 & g_{m2} & \cdots & g_{mn} \end{pmatrix}$ and so to the form $\begin{pmatrix} 1 & 0 & \cdots & 0 \\ 0 & g_{22} & \cdots & g_{2n} \\ \vdots & \vdots & & \vdots \\ 0 & g_{m2} & \cdots & g_{mn} \end{pmatrix}$ by EC(ii). By induction hypothesis, we obtain $\hat{M} = (\delta_{ij})_{r\times r}$ from M'' as required.

Definition(3.4.4). By the rank of a matrix $M = (f_{ij})_{m\times n}$, we mean the number r in the proof of proposition 3.4.3.

Example(3.4.5). We want to compute the rank of the matrix $\begin{pmatrix} 4 & 2 & 1 \\ 0 & -1 & 3 \end{pmatrix}$. Evidently its rank is not more than 2. However if we consider the linear (in)dependency of the first two columns, we come to a system of linear equations of the form

$$\begin{pmatrix} 4 \\ 0 \end{pmatrix} x + \begin{pmatrix} 2 \\ -1 \end{pmatrix} y = 0, \text{ i.e. } \begin{cases} 4x + 2y = 0 \\ \qquad -y = 0. \end{cases}$$

Since we get $x = y = 0$, it follows that the two columns are linearly independent. Hence the rank of the given matrix is just 2.

Corollary(3.4.6). *Let $M = (f_{ij})_{\ell\times\ell}$ be a square matrix. Then $D(M) \neq 0$ if and only if $M^1, \cdots, M^\ell$ are linearly independent (resp. $M_1, \cdots, M_\ell$).*

Proof. By the proof of proposition 3.4.3, M reduces to the form $M' = (\delta_{ij})_{\ell\times\ell}$ without changing the rank. Hence our assertion is clear.

Exercises of § 3-4

[1] Show that rank $(M + M') \leq$ rank $M+$ rank M' if M, M' are $m \times n$ matrices. (Hint : use mathematical induction.)

[2] What are the ranks of the following matrices ?

$$\text{(i)} \begin{pmatrix} 7 & 7 & 3 & -2 \\ -5 & -1 & 2 & 3 \\ 11 & 1 & -1 & 11 \\ 12 & 3 & 2 & 8 \end{pmatrix} \quad \text{(ii)} \begin{pmatrix} 3 & 5 & -2 \\ 2 & -1 & 3 \\ 7 & 2 & 1 \end{pmatrix}$$

$$\text{(iii)} \begin{pmatrix} 3 & 5 & -2 & 4 \\ 2 & -1 & 3 & 7 \\ 7 & 2 & 1 & 9 \end{pmatrix} \quad \text{(iv)} \begin{pmatrix} 1 & -2 & 3 & -4 \\ 5 & 6 & 7 & -8 \end{pmatrix}.$$

CHAPTER 4. RELATION OF LINEAR MAPS AND MATRICES

For a given basis, every square matrix may be recognized as a certain important mapping from a vector space to itself and vice versa. Such a fact may be futher generalized to any matrix.

§ 4-1. Linear maps

Suppose that V and W are vector spaces over a field F. We may think of a certain mapping from V into W preserving vector space structures as follows.

Definition(4.1.1). (i) A mapping L from V into W is called a *linear mapping* $L : V \to W$ if L satisfies

$$(a)\forall f \in F,\ \forall v \in V,\ L(fv) = f \cdot L(v),$$

$$(b)\forall v_1, v_2 \in V,\ L(v_1 + v_2) = L(v_1) + L(v_2).$$

Specifically we sometimes say that L is F-linear if there is a danger of confusion of the ground field. If $L(v) = 0\ \forall v \in V$, we call L a zero map.

(ii)$L^{-1}(\{0\}) = \{v \in V \mid L(v) = 0\}$ is called *the kernel* of L and is simply written by $kerL$. On the other hand, $L(V) =$

Typeset by $\mathcal{A}\mathcal{M}\mathcal{S}$-TEX

$\{L(v) \mid \forall v \in V\}$ is called *the image* of L and simply write it as ImL. If L is bijective, then the inverse mapping is easily shown to be also linear and we write it as $L^{-1} : W \to V$. In this case we also say L is invertible or L is an isomorphism. $V \cong W$ is also used as isomorphic relation. Any bijective linear mapping is obviously an isomorphism.

Proposition(4.1.2). *Let notations be as above. Then ImL and $KerL$ are also vector spaces as subspaces of W and V respectively.*

Proof. $\forall f \in F, \forall w \in ImL$, we have to check if $fw \in ImL$ first. Let $L(v) = w$; we then have $fw = fL(v) = L(fv)$ by definition of linear mapping. So $f \cdot w \in ImL$. Next if $v \in KerL$, then we must check that $fv \in KerL\ \forall f \in F$. But we see easily that $L(fv) = fL(v) = f \cdot 0 = 0$. Hence $fv \in KerL$.

Now let $w_1, w_2 \in ImL$, then $w_1 + w_2 \in ImL$. Because if $L(v_1) = w_1$ and $L(v_2) = w_2$, then $w_1 + w_2 = L(v_1) + L(v_2) = L(v_1 + v_2) \in ImL$. On the other hand $\forall v_1, v_2 \in KerL$, we have $L(v_1+v_2) = L(v_1)+L(v_2) = 0+0 = 0$. So $v_1+v_2 \in KerL$. Hence both ImL and $KerL$ become subspaces of W and V respectively.

Definition(4.1.3). A linear map $L : V \to V$ from an F-vector space into itself is called *an operator* sometimes. An operator L satisfying $L(v) = v \forall v \in V$ is called *the identity mapping* which is evidently a linear mapping and it is denoted by *id* or I. If $L(v) = \mathbb{O}\ \forall v \in V$, we then call L a trivial map, where L is any linear map having this property. In this case we write $L = 0$.

Example(4.1.4). Consider the usual scalar product in ${}_t(F^3)$ and let $M = (3, 2, -1)$. Define a mapping $L_M : {}_t(F^3) \to F$ via

$$L_M(X) = MX(\text{matrix multiplication})$$
$$M \cdot {}_tX(\text{scalar product}).$$

We may then easily perceive that L_M is just F-linear. Actually, for any $X, X' \in {}_t(F^3)$ and $\forall f \in F$, we have $L_M(X + X') = M(X + X') = MX + MX' = L_M(X) + L_M(X'); L_M(fX) = M(fX) = f(MX) = fL_M(X)$.

Example(4.1.5). Suppose that M is any $m \times n$ matrix and define a mapping $L_M : {}_tF^n \to {}_tF^m$ by $L_M(X) = MX$ $\forall$ column vector $X \in {}_tF^n$. By analogy with the above argument we are sure that L_M becomes a linear mapping.

we usually call such a mapping L_M a *linear mapping associated with* the $m \times n$ matrix M.

Remark(4.1.6). We consider the collection of all linear mappings from V into W, where V is an n-dimensional vector space and W is an m-dimensional vector space over F. Obviously this collection becomes another vector space under the sum and scalar multiplication as follows : For any such linear mappings L, L' and for any scalar $f \in F$, we define $(L+L')(v) = L(v)+L'(v)$ $\forall v \in V$ and $(fL)(v) = f\{L(v)\}$. We shall write this collection as $LM(V, W)$ called the *(vector) space of linear maps.*

Example(4.1.7). Suppose that a vector space V has a basis $\{v_1, \cdots, v_\ell\}$ and $w_1, \cdots, w_\ell$ are any elements of another vector space W. Then we may find a unique linear map L s.t. $L(v_i) = w_i \; \forall i \in \{1, \cdots, \ell\}$. Indeed $\forall f_i \in F, L(\sum_{i=1}^{\ell} f_i v_i) = \sum_{i=1}^{\ell} f_i L(v_i) = \sum_{i=1}^{\ell} f_i w_i$ holds, so that the uniqueness of such a linear map, if it exists, is obvious.

Next we have to show that L is in fact linear. $\forall v, v' \in V$, let $v = \sum_{i=1}^{\ell} f_i v_i$ and $v' = \sum_{i=1}^{\ell} f_i' v_i$. Then

$$\begin{aligned} L(v+v') &= L(\sum_{i=1}^{\ell} f_i v_i + \sum_{i=1}^{\ell} f_i' v_i) = L(\sum_{i=1}^{\ell} (f_i + f_i') v_i) \\ &= \sum_{i=1}^{\ell} (f_i + f_i') L(v_i) = \sum_{i=1}^{\ell} f_i L(v_i) + \sum_{i=1}^{\ell} f_i' L(v_i) \\ &= L(\sum_{i=1}^{\ell} f_i v_i) + L(\sum_{i=1}^{\ell} f_i' v_i) = L(v) + L(v'). \end{aligned}$$

On the other hand $\forall f \in F$,

$$\begin{aligned} L(fv) &= L(f \sum_{i=1}^{\ell} f_i v_i) = L(\sum_{i=1}^{\ell} f f_i v_i) \\ &= \sum_{i=1}^{\ell} L(f f_i v_i) = \sum_{i=1}^{\ell} f L(f_i v_i) = f L(\sum_{i=1}^{\ell} f_i v_i) \\ &= f L(v). \end{aligned}$$

Proposition(4.1.8). *Suppose that $L : V \to W$ and $L' : W \to X$ be F-linear maps for arbitrary vector spaces V, W and X. Then the composite map $L' \circ L$ also becomes linear.*

Proof. Let $v_1, v_2 \in V$ and $f_1, f_2 \in F$; then $L' \circ L(f_1v_1 + f_2v_2) = L'(f_1L(v_1) + f_2L(v_2)) = f_1L' \circ L(v_1) + f_2L' \circ L(v_2)$, so that our assertion is evident.

Corollary(4.1.9). *The following hold:*

(i) If $L_1, L_2 : V \to W$ and $L' : W \to X$ are F-linear maps, then $L' \circ (L_1 + L_2) = L' \circ L_1 + L' \circ L_2$.

(ii) If $L'_1, L'_2 : W \to X$ and $L : V \to W$ are F-linear maps, then $(L'_1 + L'_2) \circ L = L'_1 \circ L + L'_2 \circ L$.

(iii) $\forall f \in F$ and $\forall \begin{pmatrix} L : V \to W \\ L' : W \to X \end{pmatrix}$ which are linear mappings, $f(L' \circ L) = (fL') \circ L = L' \circ (fL)$.

Proof. Refer to remark 4.1.6 and proposition 4.1.8.

Definition(4.1.10). Let $L : V \to V$ be an operator of an F-vector space V. We sometimes write $L^m = L \circ L \circ \cdots \circ L$($m$-times) as an m-times composite operator of L. In addition we omit the composite notation often.

We still have an important theorem in this section related to Image and Kernel of an operator $L : V \to V$.

Theorem(4.1.11). *Suppose that V is a finite dimensional vector space and $\mathcal{O}$ is an operator of V; then $dimV = dimIm\mathcal{O} + dimKer\mathcal{O}$.*

Proof. Let $\{v_1, \cdots, v_k\}$ be a basis of the subspace $Ker\mathcal{O}$ due to proposition 4.1.2. We then extend this to a basis $\{v_1, \cdots, v_k,$

$v_{k+1}, \cdots, v_\ell\}$ of V. It is obvious that $Im\mathcal{O}$ is generated by $\mathcal{O}(v_{k+1}), \cdots, \mathcal{O}(v_\ell)$. Now they are linearly independent.

Indeed, assume $f_{k+1}\mathcal{O}(v_{k+1}) + \cdots + f_\ell\mathcal{O}(v_\ell) = \mathbb{O}$ for some $f_i \in F$ with $i = k+1, \cdots, \ell$. Then it gives rise to $\mathcal{O}(f_{k+1}v_{k+1} + \cdots + f_\ell v_\ell) = \mathbb{O}$. We thus have $f_{k+1}v_{k+1} + \cdots + f_\ell v_\ell \in Ker\mathcal{O}$.

But then $f_1v_1 + \cdots + f_kv_k + (f_{k+1}v_{k+1} + \cdots + f_\ell v_\ell) = 0$ for some $f_i \in F$ with $i = 1, 2, \cdots, \ell$. Since $\{v_1, \cdots, v_\ell\}$ is a basis of V, we absolutely have $f_i = 0 \forall i = 1, \cdots, \ell$. $\therefore$ $\{\mathcal{O}(v_{k+1}), \cdots, \mathcal{O}(v_\ell)\}$ is a basis of $Im\mathcal{O}$.

Corollary(4.1.12). *We have that $dimV$ equals the dimension of the solution space in proposition 2.3.2 plus the rank of the coefficient matrix.*

Proposition(4.1.13). *Let $L : V \to W$ be a linear map from an F-vector space V into another F-vector space W. Then $KerL = \{0\}$ if and only if L is injective.*

Proof. ($\Rightarrow$) Let $L(v_1) = L(v_2)$ for some $v_1, v_2 \in V$; we then have $L(v_1 - v_2) = 0$, so that $v_1 - v_2 = 0$, i.e., $v_1 = v_2$ proving L is injective.

($\Leftarrow$) Evident in view of the definition of injectivity.

Exercises of § 4-1

[1] Let V be the space of all differentiable functions from $\mathbb{R}$ into itself.

Show then $\dfrac{d}{dx} : V \to V$ should be an operator.

[2] Let V be as above.

Show that $\dfrac{d^n}{dx^n}; V' \to V'$ is also linear if $V' \subset V$ is the space of all n-times differentiable functions.

[3] Let V be the space of all integrable functions on $\mathbb{R}$.

Verify that the integration $\int : V \to V$ turns out to be linear.

[4] Let L be an operator as in definition 4.1.10. Suppose further that L is invertible.

Show then that L^m is also invertible.

[5] Show that if an operator L of a finite demensional F-vector space V is injective, L is actually bijective. (Hint : $V(\neq 0)$ has a basis.)

[6] Show that there may exist 1-1 operator which is not onto.

§ 4-2. Geometry of linear maps

In this section we study about convex set and its image under linear mappings. In this section we restrict ourselves to $F = \mathbb{R}$, the real field.

Definition(4.2.1). Suppose that v_0 and v_1 are elements of an F-vector space V. We call the set of all points $v_0 + sv_1$ with $0 \leq s \leq 1$ the *line segment* between v_0 and $v_0 + v_1$.

Geometrically we can illustrate the above line segment in the following manner.

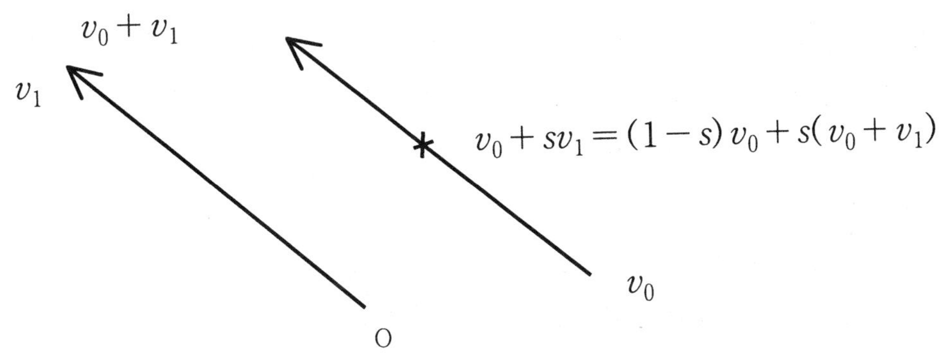

Fig. 4.2.1

Definition(4.2.2). Suppose that v_0 and v_1 are linearly independent vectors of an F-vector space V. By the *parallelogram* spanned by v_0 and v_1, we mean the set of all points designated by $s_0v_0 + s_1v_1$ with $0 \leq s_0, s_1 \leq 1$.

Noting that $s_0v_0 + s_1v_1 = (\mathbb{O} + s_0v_0) + (\mathbb{O} + s_1v_1)$, we can illustrate, using line segments, the parallelogram spanned by v_0

and v_1 in the following figure.

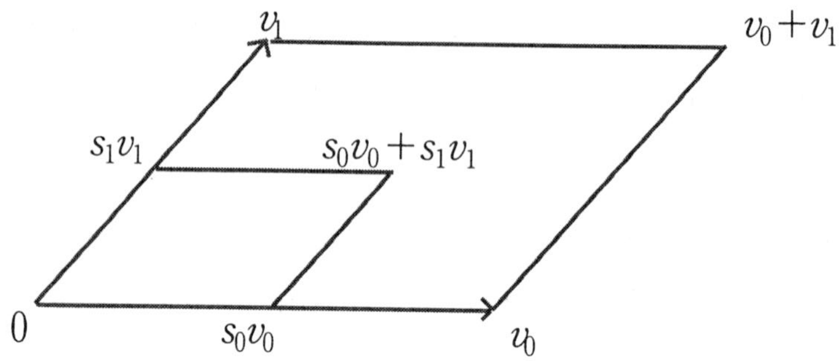

Fig. 4.2.2

Definition(4.2.3). (i) By the *translation* of a set S of V by a vector v, we mean the set $\{v+w|\forall w \in S\}$ of the vector space V.

(ii) Suppose that v_0 and v_1 are linearly independent vectors of an F-vector V. By the *triangle* spanned by $\mathbb{O}, v_0, v_1$, we mean the set $s_0v_0 + s_1v_1$ with $s_0 + s_1 \leq 1$ and $s_i \geq 0$.

(iii) Parallelogram spanned by v_0, v_1 is $\{s_0v_0+s_1v_1|0 \leq s_0, s_1 \leq 1\}$ and the translation by v of this set can by called *the most general parallelogram*, where v_0 and v_1 are independent. The most general triangle is similarly defined.

Proposition(4.2.4). *The set of all points on all line segments between v_0 and all points of line segment between $\mathbb{O}$ and v_1 is coincident with the triangle defined above.*

Proof. According to the above figures, the set of points $rv_0 + (1-r)sv_1$ with $0 \leq s, r \leq 1$ indicates the set of points between v_0 and sv_1, so that the sum of coefficients becomes $r+(1-r)s \leq$

$r + (1 - r) = 1$. So the set of points $rv_0 + (1 - r)sv_1$ is contained in the triangle defined in the preceding definition.

Next for the converse, we suppose that we are given an arbitrary point $s_0v_0 + s_1v_1$ with $s_0 + s_1 \leq 1$ and $s_i \geq 0$. Obviously $s_1 \leq 1 - s_0$. So in case $s_0 = 1$, we have $s_1 = 0$ implying our assertion. In case $s_0 < 1$, we put $r = s_0, s = \dfrac{s_1}{1 - s_0}$ and have $s_0v_0 + s_1v_1 = s_0v_0 + (1 - s_0)\dfrac{s_1}{1 - s_0}v_1 = rv_0 + (1 - r)sv_1$ showing that our converse assertion is also clear.

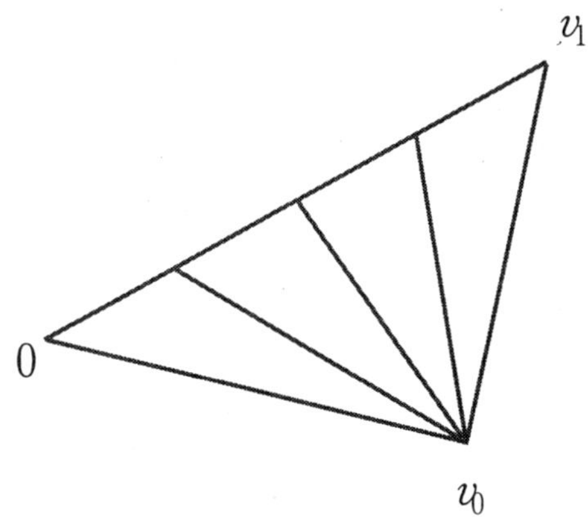

Fig. 4.2.3

Corollary(4.2.5). *Any linear map $L : V \to W$ sends a parallelogram in V to a parallelogram in W or to a line segment in W.*

Proof. By definition we may write a general parallelogram in V as $v + s_0v_0 + s_1v_1$ with $0 \leq s_0, s_1 \leq 1$ for some vectors v, v_0, v_1 in V. We thus have $L(v + s_0v_0 + s_1v_1) = L(v) + s_0L(v_0) + s_1L(v_1)$, which is just the translation of $s_0L(v_0)+s_1L(v_1)$ by $L(v)$. If $L(v_0)$

and $L(v_1)$ are linearly independent, then $s_0L(v_0) + s_1L(v_1)$ is a parallelogram spanned by $L(v_0), L(v_1)$. Otherwise it is evident that it becomes a line segment.

Definition(4.2.6). A subset S of a vector space V is called *convex* if for any given points P_0 and P_1 in S, the line segment between P_0 and P_1 is included in S.

Proposition(4.2.7). *Suppose that $P_1, \cdots, P_\ell$ are any points in a vector space V over a field F. We put $S := \{\sum_{i=1}^{\ell} s_iP_i | s_i \geq 0, \sum_{i=1}^{\ell} s_i = 1\}$. In this case S is a convex set.*

Proof. Suppose that we are given any two points P and P' designated by $P = \sum_{i=1}^{\ell} s_iP_i$, $P' = \sum_{i=1}^{\ell} s_i'P_i$, for some $s_i, s_i' \geq 0$ satisfying $\sum_{i=1}^{\ell} s_i = \sum_{i=1}^{\ell} s_i' = 1$. Now for any s with $0 \leq s \leq 1$, we may compute

$$\begin{aligned}(1-s)P + sP' &= (1-s)\sum_{i=1}^{\ell} s_iP_i + s\sum_{i=1}^{\ell} s_i'P_i \\ &= \sum_{i=1}^{\ell}\{(1-s)s_i + ss_i'\}P_i;\end{aligned}$$

but then it follows that $\forall\ i, (1-s)s_i + ss_i' \geq 0$ and

$$\begin{aligned}\sum_{i=1}^{\ell}\{(1-s)s_i + ss_i'\} &= (1-s)\sum_{i=1}^{\ell} s_i + s\sum_{i=1}^{\ell} s_i' \\ &= (1-s) + s = 1,\end{aligned}$$

which shows that S is just a convex set.

Corollary(4.2.8). *Let $L : V \to W$ be a linear map and S a subset of V defined in the preceding proposition. Then $L(S)$ becomes a convex set.*

Proof. Since S is a convex set, $L(S) = \{\sum_{i=1}^{\ell} s_i L(P_i) | s_i \geq 0, \sum_{i=1}^{\ell} s_i = 1\}$ also becomes a convex set by the preceding proposition.

Example(4.2.9). Suppose that $L : V \to \mathbb{R}$ is a linear map from an $\mathbb{R}$-vector space V into the real field $\mathbb{R}$. We claim here that $S := \{v \in V | L(v) \geq 0\}$ is a convex set in V:

For, we let $L(v_1) \geq 0$ and $L(v_2) \geq 0$; we consider $L(sv_1 + (1 - s)v_2)$ with $0 \leq s \leq 1$. We have $L(sv_1 + (1-s)v_2) = sL(v_1) + (1-s)L(v_2) \geq 0$; so that S is convex by definition.

Exercises of § 4-2

[1] Suppose that S' is a convex set in a vector space V' and L is a linear map $L : V \to V'$ from another vector space V into V'.

Show that $L^{-1}(S') = \{v|L(v) \in S'\}$ is also convex.

[2] Let S_1 and S_2 be convex subsets of a vector space V.

Show then that $S_1 \cap S_2$ is also a convex set.

[3] Let $L : V \to \mathbb{R}$ be a linear map.

Show that $L^{-1}(\mathbb{Z}^{<0}) = \{v \in V|L(v) < 0\}$ becomes a convex set.

[4] Show that the set S in proposition 4.2.7 is a convex polyhedron determined by vertices $P_1, P_2, \cdots, P_\ell$.

§ 4-3. Correspondence of matrices and linear maps

We have seen in §4-1 that any $m \times n$ matrix M induces a linear map $L_M : {}_tF^n \to {}_tF^m$ defined by $L_M(X) = MX$ $\forall$ column vector $X \in {}_tF^n$. Here we are concerned with the converse question if any linear map L from an n-dimensional vector space V into an m-dimensional vector space W may be represented isomorphically as L_M for some matrix M.

Proposition(4.3.1). *Suppose that $L : {}_tF^n \to {}_tF^m$ is a linear map ; then we have a unique matrix M satisfying $L_M = L$.*

Proof. We put $U_i's$ as standard unit vectors in F^n and $u_i's$ as standard unit vectors in F^m. Note that any vector X in ${}_tF^n$ may be written as a linear combination like

$$X = \begin{pmatrix} f_1 \\ \vdots \\ f_n \end{pmatrix} = f_1 \cdot {}_tU_1 + \cdots + f_n \cdot {}_tU_n$$

for some i-th component f_i of X in F. Because of linearity of L, we have $L(X) = L(\sum_{i=1}^n f_i \cdot {}_tU_i) = \sum_{i=1}^n f_i \cdot L({}_tU_i)$. On the other hand $L({}_tU_i)$ may be expressed as a linear combination of $u_i's$ like

$$\begin{aligned} L({}_tU_1) &= a_{11} \cdot {}_tu_1 + \cdots + a_{m1} \cdot {}_tu_m \\ &\vdots \\ L({}_tU_n) &= a_{1n} \cdot {}_tu_1 + \cdots + a_{mn} \cdot {}_tu_m \end{aligned}$$

for some numbers $a_{ij} \in F$, i.e.,

$$L({}_tU_1) = \begin{pmatrix} a_{11} \\ \vdots \\ a_{m1} \end{pmatrix}, \cdots, \ L({}_tU_n) = \begin{pmatrix} a_{1n} \\ \vdots \\ a_{mn} \end{pmatrix}$$

as column vectors. We thus have

$$L(X) = \sum_{i=1}^{n} f_i L({}_tU_i) = \sum_{i=1}^{n} f_i \begin{pmatrix} a_{1i} \\ \vdots \\ a_{mi} \end{pmatrix}$$

$$= {}_t(\sum_{i=1}^{n} a_{1i} f_i, \cdots, \sum_{i=1}^{n} a_{mi} f_i)$$

$$= \begin{pmatrix} a_{11} & \cdots & a_{1n} \\ \vdots & & \vdots \\ a_{m1} & \cdots & a_{mn} \end{pmatrix} \begin{pmatrix} f_1 \\ \vdots \\ f_n \end{pmatrix} = L_M(X),$$

where $(a_{ij})_{m \times n} = M$.

Corollary(4.3.2). *(i) If M, M' are $m \times n$ matrices, $L_{M+M'} = L_M + L_{M'}$;*

(ii) $L_{fM} = fL_M \ \forall f \in F$;

(iii) If M is an $m \times n$ matrix and M' is an $n \times s$ matrix, then $L_{MM'} = L_M \circ L_{M'}$ holds.

Proof. (i) For any column vector X with n entries,

$$L_{M+M'}(X) = (M + M')X = MX + M'X$$
$$= L_M(X) + L_{M'}(X) = (L_M + L_{M'})(X).$$

$\therefore \ L_{M+M'} = L_M + L_{M'}$;

(ii)

$$L_{fM}(X) = (fM)(X) = f(MX)$$
$$= fL_M(X) = (fL_M)(X).$$

$\therefore \ L_{fM} = fL_M$.

(iii) For any column vector X with s entries, we have

$$L_{MM'}(X) = (MM')X = M(M'X) = L_M \circ L_{M'}(X).$$

$\therefore \ L_{MM'} = L_M \circ L_{M'}$.

Proposition(4.3.3). *Suppose that V, V' are F-vector spaces with respective basis $\mathcal{B} = \{v_1, \cdots, v_n\}$ and $\mathcal{B}' = \{v'_1, \cdots, v'_m\}$. If $L : V \longrightarrow V'$ is a linear map and X is the column coordinate vector of a vector $v \in V$ relative to $\mathcal{B}$, then we have a unique*

$m \times n$ matrix M such that MX is the column coordinate vector of $L(v)$ relative to $\mathcal{B}'$. We write $M = A^{\mathcal{B}}_{\mathcal{B}'}(L)$ in this book.

Proof. Let v be expressed uniquely like $v = \sum_{i=1}^{n} f_i v_i$ for some $f_i \in F$ as a linear combination of $v_i' s$. So the map $V \longrightarrow F^n$ via $\sum f_i v_i \mapsto (f_1, \cdots, f_n)$ becomes an isomorphism. Similarly $V' \longrightarrow F^m$ via $\sum f_i' v_i' \mapsto (f_1', \cdots, f_m')$ becomes another isomorphism. So a matrix is associated with L relative to bases $\mathcal{B}, \mathcal{B}'$ according to proposition 4.3.1.

Definition(4.3.4). For a vector space V with a basis $\mathcal{B}$, $X_{\mathcal{B}}(v)$ denotes the column coordinate vector of a vector $v \in V$ relative to $\mathcal{B}$. Furthermore a unique $m \times n$ matrix M yielded as in proposition 4.3.3 shall be denoted by $A^{\mathcal{B}}_{\mathcal{B}'}(L)$ called the *associated matrix with L relative to bases $\mathcal{B}, \mathcal{B}'$.*

Remark(4.3.5). In terms of the above notations, proposition 4.3.3 shows us the formula

$$A^{\mathcal{B}}_{\mathcal{B}'}(L) \cdot X_{\mathcal{B}}(v) = X_{\mathcal{B}'}(L(v)), \text{ and}$$
$$A^{\mathcal{B}}_{\mathcal{B}'}(id) \cdot X_{\mathcal{B}}(v) = X_{\mathcal{B}'}(v)$$

holds in particular. Furthermore if we put $M := A^{\mathcal{B}}_{\mathcal{B}'}(L)$, we may write $L(v) = (M_1 \cdot X_{\mathcal{B}}(v))v_1' + \cdots + (M_m \cdot X_{\mathcal{B}}(v))v_m'$ by virtue of the proposition.

Proposition(4.3.6). *Under the notations as above, we further let L_1, L_2 be two linear maps of V into V' with respective basis*

$\mathcal{B}, \mathcal{B}'$. Let f be any number ; then we have $A^{\mathcal{B}}_{\mathcal{B}'}(L_1 + L_2) = A^{\mathcal{B}}_{\mathcal{B}'}(L_1) + A^{\mathcal{B}}_{\mathcal{B}'}(L_2)$ and $A^{\mathcal{B}}_{\mathcal{B}'}(fL_1) = f`A^{\mathcal{B}}_{\mathcal{B}'}(L_1)$.

Proof. Straightforward from corollary 4.3.2 and proposition 4.3.3.

Corollary(4.3.7). *A mapping from $LM(V, V')$ into $M_{m\times n}(F)$ via $L \mapsto A^{\mathcal{B}}_{\mathcal{B}'}(L)$ becomes an isomorphism under notations as above, where $M_{m\times n}(F)$ is the space of all $m \times n$ matrices over F.*

Proof. By virtue of proposition 4.3.6, such a mapping is clearly linear. Furthermore by virtue of example 4.1.7, such a mapping becomes onto and in addition one to one.

Proposition(4.3.8). *Suppose that vector spaces V_1, V_2, V_3 have their respective basis $\mathcal{B}, \mathcal{B}', \mathcal{B}''$. If $L_1 : V_1 \longrightarrow V_2$ and $L_2 : V_2 \longrightarrow V_3$ are linear mappings, then we obtain*

$$A^{\mathcal{B}}_{\mathcal{B}''}(L_2 \circ L_1) = A^{\mathcal{B}'}_{\mathcal{B}''}(L_2) \cdot A^{\mathcal{B}}_{\mathcal{B}'}(L_1).$$

Proof. If an arbitrary vector $v \in V_1$ has X as its column coordinate vector relative to $\mathcal{B}$, we then have $A^{\mathcal{B}}_{\mathcal{B}'}(L_1)X$ as the coordinate vector of $L_1(v)$ relative to $\mathcal{B}'$.

It follows that $A^{\mathcal{B}'}_{\mathcal{B}''}(L_2)\{A^{\mathcal{B}}_{\mathcal{B}'}(L_1)X\}$ is the column coordinate vector of $L_2(L_1(v))$ with respect to $\mathcal{B}''$. So after all we have $A^{\mathcal{B}}_{\mathcal{B}''}(L_2 \circ L_1) = A^{\mathcal{B}'}_{\mathcal{B}''} \cdot A^{\mathcal{B}}_{\mathcal{B}'}$ as required.

Corollary(4.3.9). *Suppose that $\mathcal{B}, \mathcal{B}'$ are bases of a vector space V and L is an operator of V. We then have an invertible matrix M satisfying $A^{\mathcal{B}}_{\mathcal{B}}(L) = M^{-1}A^{\mathcal{B}'}_{\mathcal{B}'}(L)M$.*

Proof. By virtue of the proposition 4.3.8, we get

$$A^{\mathcal{B}}_{\mathcal{B}}(L) = A^{\mathcal{B}'}_{\mathcal{B}}(id) \cdot A^{\mathcal{B}'}_{\mathcal{B}'}(L) \cdot A^{\mathcal{B}}_{\mathcal{B}'}(id),$$

which shows that $M = A^{\mathcal{B}}_{\mathcal{B}'}(id)$ because $A^{\mathcal{B}'}_{\mathcal{B}}(id) \cdot A^{\mathcal{B}}_{\mathcal{B}'}(id) = A^{\mathcal{B}}_{\mathcal{B}}(id)$ and so $A^{\mathcal{B}'}_{\mathcal{B}}(id) = A^{\mathcal{B}}_{\mathcal{B}'}(id)^{-1}$.

Definition(4.3.10). With notations as above if the matrix associated with L with respect to $\mathcal{B}$ is a diagonal matrix, we say that a basis $\mathcal{B}$ *diagonalize* L. We also say that L is *diagonalizable* in the sense that there may or may not exist such a basis diagonalizing L. We may also say that an $m \times m$ matrix M can be *diagonalized* over the ground field F under consideration if M considered as a linear map on V is diagonalizable.

Definition(4.3.11). An $m \times m$ matrix M_1 is said to be *similar* to another $m \times m$ matrix M_2 in case we have an invertible matrix M_3 satisfying $M_2 = M_3^{-1}M_1M_3$; we may also say that two matrices M_1, M_2 are *similar* over F.

Exercises of § 4-3

[1] Let $\mathcal{B}$ be a basis of an m-dimensional vector space V over F; show that $A_{\mathcal{B}}^{\mathcal{B}}(id) = I_m$.

[2] Let V_1, V_2 are vector spaces over $F = \mathbb{R}$ with $dimV_1 = 2, dimV_2 = 3$. Suppose that we have a linear map $L : V_1 \longrightarrow V_2$ defined by $L(v_{11}) = 2v_{21} + 4v_{22} - 3v_{23}$, $L(v_{12}) = -v_{21} + 7v_{22} + 8v_{23}$, where $\mathcal{B}_1 = \{v_{11}, v_{12}\}$ is a basis of V_1 and $\mathcal{B}_2 = \{v_{21}, v_{22}, v_{23}\}$ is a basis of V_2. In this case compute $A_{\mathcal{B}_2}^{\mathcal{B}_1}(L)$ over $F = \mathbb{R}$.

[3] Let $L : \mathbb{R}^2 \longrightarrow \mathbb{R}^2$ be a linear map satisfying $A_{\mathcal{B}}^{\mathcal{B}}(L) = \begin{pmatrix} cos\theta & -sin\theta \\ sin\theta & cos\theta \end{pmatrix}$, where $\mathcal{B} = \{U_1, U_2\}$ is the standard basis of $\mathbb{R}^2$. Verify that L is a rotation by an angle θ.

[4] Let V, V' be vector spaces over F with respective basis $\mathcal{B} = \{v_1, \cdots, v_n\}$, $\mathcal{B}' = \{v'_1, \cdots, v'_n\}$.

Let $v'_i = \sum_{j=1}^{n} f_{ij} v_j$ for some $f_{ij} \in F$ with $1 \leq i, j \leq n$. Suppose that L is a linear map defined by $L(v_1) = v'_1, \cdots, L(v_n) = v'_n$. Compute $A_{\mathcal{B}}^{\mathcal{B}}(L)$ and $A_{\mathcal{B}}^{\mathcal{B}'}(id)$.

[5] Show that any projection $\pi_i : F^m \longrightarrow F$ defined by $(f_1, \cdots, f_m) \mapsto f_i$ for $1 \leq i \leq m$ becomes a linear mapping.

[6] Prove proposition 4.3.6 in detail.

CHAPTER 5. BILINEAR MAPPINGS

Scalar product is a special example of a general concept, bilinear mapping. In this chapter we mainly learn about its concept and its associated matrix which is similar in spirit to the one in the preceding section if the codomain is one-dimensional.

§ 5-1. Orthogonal basis

For the purpose of discussing about othogonal basis of particular vector spaces we need to define the following at first.

Definition(5.1.1). For given 3 vector spaces V_1, V_2, V_3 over F, let $\beta : V_1 \times V_2 \longrightarrow V_3$ be a map satisfying that $(v_1, v_2) \mapsto \beta(v_1, v_2)$ is linear with respect to v_1 and v_2 respectively. In this case we call β *a bilinear mapping.* In particular if $V_3 = F$, we call β *a bilinear form.*

More precisely for any $v_1, v_2 \in V_1; v_1', v_2' \in V_2; f_1, f_2 \in F$, β should satisfy $\beta(f_1 v_1 + f_2 v_2, v') = f_1 \beta(v_1, v') + f_2 \beta(v_2, v')$ $\forall v' \in V_2$ and similarly $\beta(v, f_1 v_1' + f_2 v_2') = f_1 \beta(v, v_1') + f_2 \beta(v, v_2')$ $\forall v \in V_1$.

Example(5.1.2). Suppose that for a given matrix $M = (f_{ij})_{m\times n}, \beta_M : F^m \times {}_tF^n \longrightarrow F$ is defined by $\beta_M(X, X') = XMX'$. If $X = (x_1, \cdots, x_m), X' = {}_t(x'_1, \cdots, x'_n)$ are desingated, we should obtain $\beta_M(X, X')$ more speciafically as

$$(x_1, \cdots, x_m)\begin{pmatrix} f_{11} & \cdots & f_{1n} \\ \vdots & & \vdots \\ f_{m1} & \cdots & f_{mn} \end{pmatrix}\begin{pmatrix} x'_1 \\ \vdots \\ x'_n \end{pmatrix}$$

in terms of matrix multiplication. We should like to check if β_M is a bilinear map.

Suppose that X_1, X_2 are any row vectors in F^m and f_1, f_2 are any scalars in F; we then have : $\forall X' \in {}_tF^n$,

$$\begin{aligned} &\beta_M(f_1X_1 + f_2X_2, X') \\ &= (f_1X_1 + f_2X_2)MX' \\ &= f_1X_1MX' + f_2X_2MX' \\ &= f_1\beta_M(X_1, X') + f_2\beta_M(X_2, X'). \end{aligned}$$

similarly we have

$$\begin{aligned} &\beta_M(X, f_1X'_1 + f_2X'_2) \\ &= f_1\beta_M(X, X'_1) + f_2\beta_M(X, X'_2) \\ &\forall X \in F^m; \forall X'_1, X'_2 \in {}_t(F^n); \forall f_1, f_2 \in F. \end{aligned}$$

So β_M is a bilinear map. Furthermore

$$\begin{aligned}\beta_M(X, X') &= XMX' = (XM)X' \\ &= (\sum_{i=1}^{m} x_i f_{i1}, \cdots, \sum_{i=1}^{m} x_i f_{in}) \cdot X' \\ &= \sum_{j=1}^{n} \sum_{i=1}^{m} x_i f_{ij} x'_j \\ &= \sum_{j=1}^{n} \sum_{i=1}^{m} f_{ij} x_i x'_j\end{aligned}$$

is obtained as an element of F.

Proposition(5.1.3). *Suppose that we are given a bilinear form $\beta : F^m \times {}_tF^n \longrightarrow F$; then we have a unique matrix M satisfying $\beta(X, X') = XMX'$ $\forall X \in F^m, \forall X' \in {}_tF^n$. M is called the matrix representing β.*

Proof. Suppose that $U_1, \cdots, U_m$ are the standard unit vectors of F^m and $U^1, \cdots, U^n$ are the standard unit vectors of ${}_tF^n$. For any $X \in F^m$ and $X' \in {}_t(F^n)$, we may put $X := \sum_{i=1}^m f_i U_i$ and $X' := \sum_{j=1}^n f'_j U^j$ with $f_i, f'_j \in F$. But then

$$\begin{aligned}&\beta(X, X') \\ &= \beta(\sum_{i=1}^{m} f_i U_i, \sum_{j=1}^{n} f'_j U^j) \\ &= \sum_{i=1}^{m} f_i \beta(U_i, \sum_{j=1}^{n} f'_j U^j) \\ &= \sum_{i=1}^{m} \sum_{j=1}^{n} f_i f'_j \beta(U_i, U^j)\end{aligned}$$

is obtained. So if we let $M := (f_{ij})$ with $f_{ij} = \beta(U_i, U^j)$, we may have

$$\begin{aligned}\beta(X, X') &= \sum_{i=1}^{m}\sum_{j=1}^{n} f_i f'_j \beta(U_i, U^j) \\ &= XMX' \\ &= (f_1, \cdots, f_m)M\begin{pmatrix} f'_1 \\ \vdots \\ f'_n \end{pmatrix} \\ &= (f_1, \cdots, f_m)M\,{}_t(f'_1, \cdots, f'_n)\end{aligned}$$

as required. Next we must show that such a matrix M is unique. Suppose that we have another matrix N so that $XMX' = XNX'$; but then we get $XMX' - XNX' = X(M-N)X' = 0$ for all $X \in F^m$ and for all $X' \in {}_tF^n$. If we put $A := M - N = (a_{ij})$, then we have $XAX' = 0$. Let $X = U_i$ and $X' = U^j$; then we get $XAX' = a_{ij} = 0$, which follows for all i, j such that $1 \leq i \leq m$ and $1 \leq j \leq n$. We conclude that $M - N = 0$, so that $M = N$.

Example(5.1.4). Let $\beta : F^2 \times {}_tF^3 \longrightarrow F$ be a bilinear map defined by $\beta(X, X') = XMX'$, where $M = \begin{pmatrix} 2 & 3 & -4 \\ 5 & 1 & 4 \end{pmatrix}$. If we put $X = (f_1, f_2)$, $X' = {}_t(f'_1, f'_2, f'_3)$, then it follows that

$$\begin{aligned}XMX' &= (f_1, f_2)\begin{pmatrix} 2 & 3 & -4 \\ 5 & 1 & 4 \end{pmatrix}\begin{pmatrix} f'_1 \\ f'_2 \\ f'_3 \end{pmatrix} \\ &= (2f_1 + 5f_2, 3f_1 + f_2, -4f_1 + 4f_2)\begin{pmatrix} f'_1 \\ f'_2 \\ f'_3 \end{pmatrix} \\ &= (2f_1 + 5f_2)f'_1 + (3f_1 + f_2)f'_2 + (-4f_1 + 4f_2)f'_3.\end{aligned}$$

Example(5.1.5). If we let $M = \begin{pmatrix} 1 & 2 \\ -3 & -1 \end{pmatrix}$ and if $X = (f_1, f_2)$ and $X' = {}_t(f_1', f_2')$ are adopted, then

$$\begin{aligned} XMX' &= (f_1, f_2)\begin{pmatrix} 1 & 2 \\ -3 & -1 \end{pmatrix}\begin{pmatrix} f_1' \\ f_2' \end{pmatrix} \\ &= (f_1 - 3f_2, 2f_1 - f_2)\begin{pmatrix} f_1' \\ f_2' \end{pmatrix} \\ &= (f_1 - 3f_2)f_1' + (2f_1 - f_2)f_2' \\ &= f_1f_1' + 2f_1f_2' - 3f_2f_1' - f_2f_2'. \end{aligned}$$

We now focus on orthogonal basis of a vector space. As is easily seen, scalar product is a special case of a general notion as above. Suppose that we have a finite dimensional vector space V over a field $F = \mathbb{R}$. We need to define some more definitions for our purport.

Definition(5.1.6). A bilinear form $\beta : V \times V \longrightarrow F$ is called *symmetric* if $\beta(v, w) = \beta(w, v)$ for all $v, w \in V$. A symmetric bilinear form is called a *scalar product* generally and is denoted by $<,>$ sometimes.

Proposition(5.1.7). *A symmetric bilinear form on $F^n \times {}_tF^n$ is represented by an $n \times n$ matrix in F if and only if it is a symmetric matrix.*

Proof. (**necessity**) Assume that an $n \times n$ matrix M represents a symmetric bilinear form, namely $XMX' = {}_tX'M_tX \forall X \in F^n, \forall X' \in {}_tF^n$. Because ${}_t({}_tX'M_tX) = X_tMX' = XMX'$, i.e. $X(M - {}_tM)X' = 0$, we have $M = {}_tM$.

(**sufficiency**) Suppose that M is an $n \times n$ symmetric matrix. By virtue of example 5.1.2, M defines a bilinear map $F^n \times {}_tF^n \longrightarrow F$. But we know easily that $XMX' = {}_t(XMX') = {}_tX'{}_tM{}_tX' = {}_tX'M{}_tX \ \forall \in F^n, \forall X' \in {}_tF^n$, showing that M represents a symmetric bilinear form.

Remark(5.1.8). Suppose that we are given a bilinear form $\beta : V \times V \longrightarrow F$ for a finite dimensional vector space V over F with a basis $\{v_1, \cdots, v_n\}$ of V. Let $v = \sum_{i=1}^{\ell} f_i v_i$ and $v' = \sum_{i=1}^{\ell} f_i' v_i$ with $f_i, f_i' \in F$. We then have $\beta(v, v') =, \sum_{i,j=1}^{n} f_i f_j' \beta(v_i, v_j)$. So if we put $M := (f_{ij})$ with $f_{ij} = \beta(v_i, v_j)$, it follows that $\beta(v, v') = \sum_{i,j=1}^{n} f_i f_j' f_{ij} = XMX'$, where $X = (f_i) \in F^n$ and $X' = {}_t(f_j') \in {}_tF^n$. We shall call M the *matrix representing the bilinear form* β *on* V.

Proposition(5.1.9). *If we denote the set of all bilinear maps of* $F^m \times F^n$ *into* F *by* $BM(F^m \times F^n, F)$*, then we have an isomorphism between* $M_{m\times n}(F)$ *and* $BM(F^m \times F^n, F)$*.*

Proof. Evidently $BM(F^m \times F^n, F)$ is a vector space over F. We first give the association $M \mapsto \beta_M$ from $M_{m\times n}(F)$ into $BM(F^m \times F^n, F)$. By virtue of example 5.1.2 and proposition 5.1.3, this is a well defined map, which is actually an isomorphism as is easily seen.

Definition(5.1.10). (i) If a scalar product $<,>$ on V satisfies that $\exists v \in V$ such that $< v, v' >= 0 \ \forall v' \in V$ implies $v = 0$, then we say the scalar product $<,>$ is *nondegenerate*.

(ii) If $< v, v' >= 0$ for some elements $v, v' \in V$ with a scalar product $<,>$, we say that v and v' are *perpendicular or orthogonal* and write $v \perp v'$. For a subset S of V, $S^{\perp} := \{v \in V| < v, v' >= 0 \forall v' \in S\}$ is evidently a subspace of V, called the orthogonal space of S sometimes. We call a basis $\{v_1, \cdots, v_\ell\}$ *orthogonal* if $< v_i, v_j >= 0$ for any $i \neq j$.

(iii) If V is a vector space over $\mathbb{R}$ with a scalar product satisfying that $< v, v >> 0 \forall v \neq 0$, then $<,>$ is called *positive definite.* We define $\|v\| :=< v, v >^{\frac{1}{2}}$ and call it the norm or length of an element $v \in V$. If $\|v\| = 1$, we call $v \in V$ a unit vector . An orthogonal basis $\{v_1, \cdots, v_\ell\}$ is called *orthonormal* if each v_i is a unit vector.

Theorem(5.1.11). *(***Schwarz inequality***) Let V be a vector space over $\mathbb{R}$ with a positive definite scalar product. Then $| < v, v' > | \leqq \|v\|\|v'\| \forall v, v' \in V$.*

Proof. If $v' = 0$, the assertion is obvious. Otherwise $\forall r \in \mathbb{R} < v + rv', v + rv' >= r^2 < v', v' > +2r < v, v' > + < v, v >\geqq 0$. So $< v, v' >^2 - < v, v >< v', v' >\leqq 0$, i.e., $| < v, v' > |^2 \leqq \|v\|^2\|v'\|^2$ proving our assertion.

Theorem(5.1.12). *(***Triangle inequality***) With assumptions as in theorem 5.1.11, $\|v + v'\| \leqq \|v\| + \|v'\|$.*

Proof. It suffices to show that $\|v + v'\|^2 \leqq (\|v\| + \|v'\|)^2$. But using Schwarz inequality, we have $\|v+v'\|^2 =< v+v', v+v' >=< v, v > +2 < v, v' > + < v', v' >\leqq \|v\|^2 + 2\|v\|\|v'\| + \|v'\|^2 =$

$(\|v\| + \|v'\|)^2$ immediately.

Theorem(5.1.13). *(***Gram-Schmidt orthogonalization process***) With assumptions as in theorem 5.1.11, let S be a subspace of V and let $\{s_1, \cdots, s_k\}$ be an orthogonal basis of S.*

Then we have vectors $s_{k+1}, \cdots, s_\ell$ of V such that $\{s_1, \cdots, s_k, s_{k+1}, \cdots, s_\ell\}$ is an orthogonal basis of V if $S \neq V$ and V is finite dimensional.

Proof. By virtue of the knowhow in chapter 1, there exists a basis $\{s_1, \cdots, s_k, v_{k+1}, \cdots, v_\ell\}$ of V for some $v_{k+1}, \cdots, v_\ell \in V$. Let S_{k+1} be the subspace of V generated by the vectors $s_1, \cdots, s_k, v_{k+1}$ and let $c_i := \frac{<v_{k+1}, s_i>}{<s_i, s_i>}$ $\forall i \in \{1, \cdots, k\}$. If we put $s_{k+1} := v_{k+1} - \sum_{i=1}^k c_i s_i \neq 0$, we immediately see that $s_{k+1} \perp s_1, \cdots, s_{k+1} \perp s_k$. Indeed

$$\begin{aligned}
&< s_{k+1}, s_i > \\
&=< v_{k+1} - \sum_{i=1}^{k} c_i s_i, s_i > \\
&=< v_{k+1}, s_i > - < c_i s_i, s_i > \\
&=< v_{k+1}, s_i > - c_i < s_i, s_i > \\
&= 0 \ \forall i \in \{1, \cdots, k\}.
\end{aligned}$$

So S_{k+1} has an orthogonal basis $\{s_1, \cdots, s_{k+1}\}$. Proceeding in this way by mathematical induction, the subspace S_{k+j} generated by the vectors $s_1, \cdots, s_k, v_{k+1}, \cdots, v_{k+j}$ with $1 \leq j \leq \ell - k$ should have an orthogonal basis $\{s_1, \cdots, s_k, \cdots, s_{k+j}\}$.

Proposition(5.1.14). *With assumptions as in the preceding theorems, let S be a subspace of V. Then $dimS + dimS^{\perp} = dimV$ holds.*

Proof. In case $S = V$ or $S = \mathbb{O}$, the equality is trivial. So we assume $S \neq V$ and $S \neq \mathbb{O}$. Once we have an orthogonal basis by theorem 5.1.13, we may get an orthonormal basis dividing each basis element by its norm. So if $\{s_1, \cdots, s_k\}$ is an orthonormal basis of S, we may find vectors $u_{k+1}, \cdots, u_\ell$ of V so that $\{s_1, \cdots, s_k, u_{k+1}, \cdots, u_\ell\}$ is an orthonormal basis of V. We claim here that $S^{\perp}$ has an orthonormal basis $\{u_{k+1}, \cdots, u_\ell\}$.

Suppose that v is an arbitrary element of $S^{\perp}$ expressed as a linear combination like $v = f_1 s_1 + \cdots + f_k s_k + f_{k+1} u_{k+1} + \cdots + f_\ell u_\ell$ for some $f_1, \cdots, f_\ell \in F = \mathbb{R}$ here. Since v is perpendicular to S, we obtain $0 = < v, s_i > = f_i < s_i, s_i > = f_i \forall i \in \{1, \cdots, k\}$, so that v is just a linear combination of $u_{k+1}, \cdots, u_\ell$.

Finally any linear combination of $u_{k+1}, \cdots, u_\ell$ becomes orthogonal to any s_i with $i = 1, \cdots, k$. So $S^{\perp}$ is generated by $u_{k+1}, \cdots, u_\ell$, and hence $\{u_{k+1}, \cdots, u_\ell\}$ constitutes an orthonormal basis of $S^{\perp}$, completing the proof.

Corollary(5.1.15). *With assumptions as in proposition 5.1.14, $V = S \oplus S^{\perp}$ holds. $S^{\perp}$ is also called the orthogonal complement of S.*

Proof. Using proposition 1.4.4 and proposition 5.1.14 finishes the proof.

Now we'd like to do similar jobs as above with Hermitian form or equivalently Hermitian product given instead of scalar product.

Definition(5.1.16). Let V be a vector space over the field $\mathbb{C}$ of complex numbers. By a hermitian product we mean a function $<,>: V \times V \longrightarrow \mathbb{C}$ satisfying the following rules:

(i) $\forall v, v' \in V, < v, v' >=< \overline{v', v} >$, where $-$(bar) indicates complex conjugate.

(ii) $\forall v, v', v'' \in V, < v, v' + v'' >=< v, v' > + < v, v'' >$.

(iii) $\forall c \in \mathbb{C}, \forall v, v' \in V, < cv, v' >= c < v, v' >, < v, cv' >= \bar{c} < v, v' >$.

Furthermore if a hermitian product $<,>$ satisfies $< v, v >> 0 \forall v (\neq 0) \in V$, we then say that the hermitian product is *positive definite*. We may also define the terminologies *perpendicular, orthogonal, orthogonal basis, orthogonal complement* similar to the ones before. We may also define the words unit vector, orthonormal basis as before.

Example(5.1.17). For any vectors $X, Y \in \mathbb{C}^m$, we designate $< X, Y >= X \cdot {}_t\bar{Y} = \sum_{i=1}^m x_i \bar{y}_i$, where $X = (x_1, \cdots, x_m), Y = (y_1, \cdots, y_m)$. Obviously this $<,>$ is rendered a hermitian product. In addition, it becomes positive definite noting that $x \cdot \bar{x} = \|x\|^2 \ \forall x \in \mathbb{C}$.

Now we have analogues of the preceding facts, which are written down without proofs as follows.

Proposition(5.1.18). *(***Gram-Schmidt orthogonalization process***) Suppose that V is a finite dimensional vector space over $\mathbb{C}$ with a positive definite hermitian product. Let $\{s_1, \cdots, s_k\}$ be an orthogonal basis of a subspace $S \subsetneqq V$.*

Then we may find elements $s_{k+1}, \cdots, s_\ell$ satisfying $\{s_1, \cdots, s_k, s_{k+1}, \cdots, s_\ell\}$ constitutes an orthogonal basis of V.

Corollary(5.1.19). *(i) Let S be a subspace of V.*

Then $dimS + dimS^\perp = dimV$ holds.

(ii) $V = S \oplus S^\perp$ holds.

Remark(5.1.20). The positive definite hermitian product takes after the positive definite scalar product in many properties among which some are left as exercises to the reader.

Hermitian product shows up again in §7-2 because of its importance. Note however that a hermitian product is not bilinear.

Exercises of § 5-1

[1] Suppose that we are given a positive definite hermitian product on a vector space V over $\mathbb{C}$. Verify the following statements.

(i)$\forall v, v' \in V, | < v, v' > | \leq \|v\| \|v'\|$,

(ii)$\forall v \in V, \|v\| \geq 0$, and $\|v\| = 0$ if and only if $v = 0$,

(iii)$\forall c \in \mathbb{C}, \forall v \in V, \|cv\| = |c| \|v\|$,

(iv)$\forall v, v' \in V, \|v + v'\| \leq \|v\| + \|v'\|$.

[2] Prove proposition 5.1.18 in detail.

[3] Prove corollary 5.1.19 in detail.

[4] Prove or disprove (i) $\forall v, v' \in V, \|v + v'\|^2 = \|v\|^2 + \|v'\|^2$ if v and v' are perpendicular,

(ii) $\forall v, v' \in V, \|v + v'\|^2 + \|v - v'\|^2 = 2\|v\|^2 + 2\|v'\|^2$ in a vector space V with a positive definite scalar product or with a positive definite hermitian product.

[5] Let us given a vector space V of continuous complex valued functions defined on the interval $[-\pi, \pi]$. $\forall f_1, f_2 \in V$, we put $< f_1, f_2 >:= \int_{-\pi}^{\pi} f_1(x)\bar{f_2}(x)dx$. Show then that $<, >$ becomes a hermitian product which is positive definite.

[6] Let V be a vector space of continuous real valued functions defined on the interval $[-1, 1]$. $\forall f_1, f_2 \in V$, we put $< f_1, f_2 >= \int_{-1}^{1} f_1(x) f_2(x)dx$. Verify that $<, >$ is nonde-

generate. Is it positive definite scalar product in addition ?

[7] What's the difference between positive definiteness and nondegeneracy ?

§ 5-2. Dual space

Since F is a 1-dimensional vector space, we may consider the vector space $LM(V, F)$ when V is a vector space over F.

Definition(5.2.1). We denote $LM(V, F)$ simply by V^* and call it the *dual space* of V. Elements of V^* are called *linear forms* or *functionals* on V.

Example(5.2.2). If a vector space V over F has a scalar product $<,>$ then the map $v \mapsto < v, v' >$ for $\forall v \in V$ and for a fixed element $v' \in V$ is obviously a linear form.

Example(5.2.3). Suppose that a vector space V over the complex number field $\mathbb{C}$ has a hermitian product $<,>$. Although the map $v \mapsto < v, v' >$ for a fixed vector $v' \in V$ is a linear form, the map $v \mapsto < v', v >$ is not a linear form in general since $cv \mapsto < v', cv >= \bar{c} < v', v >$ holds. In view of this, the map $L : V_1 \longrightarrow V_2$ of vector spaces over a field F having conjugates is called *semi-linear* or *anti-linear* if $L(v_1 + v_1') = L(v_1) + L(v_1')$ but $L(fv_1) = \bar{f}L(v_1)$ $\forall v_1, v_1' \in V$, $\forall f \in F$, where-(bar)denotes the conjugate.

Proposition(5.2.4). *Suppose that V is a finite dimensional vector space over a field F; then V^* is finite dimensional and $dimV^* = dimV$.*

Proof. If $\{v_1, \cdots, v_\ell\}$ is a basis of V, we define functionals v_i^* as

$v_i^*(v_j) = \delta_{ij}\ \forall i, j \in \{1, \cdots, \ell\}$, i.e.,

$$\begin{cases} v_i^*(v_j) = 1 \text{ if } i = j, \\ v_i^*(v_j) = 0 \text{ if } i \neq j. \end{cases}$$

Obviously $v_1^*, \cdots, v_\ell^*$ are functionals and they are linearly independent because $\sum_{i=1}^{\ell} f_i v_i^* = 0$ for some $f_i's \in F$ implies $0 = \sum_{i=1}^{\ell} f_i v_i^*(v_j) = f_j v_j^*(v_j) = f_j \forall j = 1, \cdots, \ell$.

Next we claim that $\forall v^* \in V^*$ has the expression of the form $v^* = \sum_{i=1}^{\ell} v^*(v_i) v_i^*$. This is so because both hand sides have the same value for each basis element $v_i, i = 1, \cdots, \ell$. So we have the conclusion.

Definition(5.2.5). The basis $\{v_1^*, \cdots, v_\ell^*\}$ of V^* in the preceding proposition is called *the dual basis* of $\{v_1, \cdots, v_\ell\}$ in V.

Proposition(5.2.6). *Let V be a finite dimensional vector space over a field F with a nondegenerate scalar product $<,>$. If we denote the linear form $v \mapsto < v, w > \ \forall v \in V$ by L_w for a fixed $w \in V$, then we have an isomorphism of V into V^* via $w \mapsto L_w$.*

Proof. Since a scalar product is a symmetric bilinear form, it is obvious that the map $w \mapsto L_w$ is linear. Next suppose that $L_w = 0$ for some $w \in V$; then $\forall v \in V, 0 = L_w(v) =< w, v >$, so that $w = 0$ by the nondegeneracy of the scalar product. So in view of proposition 5.2.4, the map $w \mapsto L_w$ is an isomorphism.

Corollary(5.2.7). *Let V be a vector space as in proposition 5.2.6; then any linear form is of the form L_w for some $w \in V$.*

Proof. Straightforward fact from the isomorphism of V into V^* via $w \mapsto L_w$.

Proposition(5.2.8). *Let S be a subspace of a finite dimensional F-vector space V with a nondegenerate scalar product. We then have $dimS + dimS^{\perp} = dimV$, and moreover $S \oplus S^{\perp} = V$ if $S + S^{\perp} = V$.*

Proof. We put ${}^{\perp}S := \{x \in V^* | x(S) = 0\}$ which is evidently a subspace of V^*. Let $\{s_1, \cdots, s_k\}$ be a basis of S. By definition this basis of S may be extended to a basis $\{s_1, \cdots, s_k, \cdots, s_\ell\}$ of V. Obviously $\{s^*_{k+1}, \cdots, s^*_\ell\} \subset {}^{\perp}S$. So as to show it is actually a basis of ${}^{\perp}S$, we need to show $\{s^*_{k+1}, \cdots, s^*_\ell\}$ generates ${}^{\perp}S$. Suppose that $x \in {}^{\perp}S$ has an expression like $x = \sum_{i=1}^{\ell} f_i s_i^*$ for some $f_i \in F, i = 1, \cdots, \ell$; then $\forall s_i$ for $i = 1, \cdots, k, 0 = x(s_i) = f_i$ happens. So we have just shown $dimS + dim{}^{\perp}S = dimV$.

Finally we consider the isomorphism $w \mapsto L_w$ between $S^{\perp}$ and ${}^{\perp}S$ using proposition 5.2.6. Hence we have $dimS + dimS^{\perp} = dimV$ as asserted. Because $S \cap S^{\perp} = (0)$, evidently $V = S \oplus S^{\perp}$.

Remark(5.2.9). If a functional $L \in V^*$ is of the form L_w for some $w \in V$, we say that w represents the functional L with respect to the given nondegenerate scalar product.

Exercises of § 5-2

[1] Let V be a vector space over $\mathbb{C}$ with a hermitian product $<,>$. Verify that the map $v \mapsto < w, v >$ for a fixed $w \in V$ is semi-linear or anti-linear,i.e., $v, v' \in V, < w, v+v' >=< w, v > + < w, v' >$ and $\forall c \in C < w, cv >= \bar{c} < w, v >$.

[2] Suppose that V is a finite demensional vector space over a field F with a nondegenerate scalar product. Using proposition 5.2.8, show that $S^{\perp\perp} = S$ for any subspace S of V.

[3] Let V be the $\mathbb{R}$-vector space $\mathbb{R}^m$ with the usual dot product. If S is a subspace of V, what are dimensions of S^* and $S^\perp$ respectively ?

[4] Suppose that $p_i : F^m \longrightarrow F$ is the projection on the ith coordinate, i.e., $p_i(x_1, \cdots, x_m) = x_i \forall i \in \{1, \cdots, m\}$. Show that each p_i is a linear form on F^m.

[5] If V^{**} denotes the dual space of V^*, show that every element $v \in V$ yields an element x_v in V^{**} so that $v \mapsto x_v$ brings about an isomorphism between V and V^{**}. Here we assume V is finite dimensional.

§ 5-3. Revisit of linear equations

Suppose that we have a system of m-linear equations in n-unknowns like:

$$\begin{cases} f_{11}x_1 + \cdots + f_{1n}x_n = 0 \\ \vdots \qquad\qquad \vdots \\ f_{m1}x_1 + \cdots + f_{mn}x_n = 0, \end{cases}$$

where $f_{ij} \in F, i = 1, \cdots, m, j = 1, \cdots, n$. A nice application of previous sections to this homogeneous system happens in the following sense.

Proposition(5.3.1). *If we put $M = (f_{ij})_{m\times n}$ and $\mathbb{X} = {}_t(x_1, \cdots, x_n)$, then the solution space of the above homogeneous system $\sum_{i=1}^n x_i M^i = 0$ constitutes the space perpendicular to row vectors of M, i.e., $M_1, \cdots, M_m$, with respect to the usual scalar product on F^n.*

Proof. We consider the linear map $L_M : {}_tF^n \longrightarrow {}_tF^m$ and find that the kernel of L_M is none other than the solution space of the equation $M\mathbb{X} = \mathbb{O}$, i.e., $M_1\mathbb{X} = \cdots = M_m\mathbb{X} = \mathbb{O}$ which is equivalent to the form $\sum_{i=1}^n x_i M^i = 0$ as explained in proposition 2.3.2.

Example(5.3.2). Look at the following system of linear equations for a moment :

$$\begin{cases} 2x_1 + 2x_2 + x_3 = 1, \\ x_2 - x_3 = 0. \end{cases}$$

We may find decently the dimension of the solutions of this system mentioned in example 2.3.4. Since it has the form

$$\begin{pmatrix} 2 & 2 & 1 \\ 0 & 1 & -1 \end{pmatrix} {}_t(x_1, x_2, x_3) = \begin{pmatrix} 1 \\ 0 \end{pmatrix}$$

and since the dimension of the space S generated by M_1, M_2 of $M := \begin{pmatrix} 2 & 2 & 1 \\ 0 & 1 & -1 \end{pmatrix}$ is just the $rankM = 2$, we know $dimS + dimS^{\perp} = 3$, i.e., $dimS^{\perp} = 3 - 2 = 1$ by virtue of proposition 5.2.8. So the required dimension is 1 which was noticed before directly.

Example(5.3.3). We'd like to search a basis for the solution space S of the equation $3x_1 - 2x_2 + x_3 = 0$. By dint of proposition 5.2.8, we have the number of unknowns $= dimS + dimS^{\perp}$, i.e., $dimS = 3 - 1 = 2$. Although we may take various bases of S, we'd like to try $s_2 = (0, 1, 0), s_3 = (0, 0, 1)$ putting $s_1 = (3, -2, 1)$ and orthogonalize them using the analogue of the Gram-Schmidt orthogonalization process.

We have to show s_1, s_2, s_3 are linearly independent in the first place. For this, let $\sum_{i=1}^{3} f_i s_i = 0$ for some $f_i \in F, i = 1, 2, 3$. From corollary 3.4.6, we know $3 = D(s_1, s_2, s_3) \neq 0$, which guarantees us the linear independence of s_1, s_2, s_3. Next we put $\tilde{s_2} := s_2 - \frac{<s_2, s_1>}{<s_1, s_1>} s_1$ and $\tilde{s_3} := s_3 - \frac{<s_3, s_1>}{<s_1, s_1>} s_1 - \frac{<s_3, \tilde{s_2}>}{<\tilde{s_2}, \tilde{s_2}>} \tilde{s_2}$. Since we are dealing with an ordinary dot product, it follows that $\tilde{s_2} = (\frac{3}{7}, \frac{5}{7}, \frac{1}{7})$ and $\tilde{s_3} = (0, 0, 1) - \frac{1}{14}(3, -2, 1) - \frac{1}{35}(3, 5, 1)$. Hence the solution space of the given equation should have $\{\tilde{s_2}, \tilde{s_3}\}$ as its orthogonal basis.

Corollary(5.3.4). *Under notations in proposition 5.3.1, the solution space has its dimension $n - r$, where r is the rank of M and n is the number of unknowns of the homogeneous system of linear equations.*

Proof. Very clear by virtue of proposition 5.3.1, and proposition 5.2.8.

Exercises of § 5-3

[1] What is the dimension of the solution space of the following system of linear equations ?

(i) $\begin{cases} x_1 - x_2 + x_3 = 0 \\ x_1 + x_2 = 0 \\ x_2 - x_3 = 0. \end{cases}$ (ii) $\begin{cases} -2x_1 + x_2 + x_3 = 0 \\ 7x_1 + 4x_2 - \pi x_3 = 0. \end{cases}$

[2] Find out the dimension of the solution space of the following systems of linear equations in 3 unknowns.

(i) $\begin{cases} x_1 - x_2 + x_3 = 0 \\ 4x_1 - 4x_2 + 4x_3 = 0. \end{cases}$ (ii) $3x_1 + 4x_2 - 5x_3 = 0.$

[3] For a given matrix $M = (f_{ij})_{m\times n}$, we let $L_M : {}_tF^n \longrightarrow {}_tF^m$ be the linear map defined in example 4.1.5. Verify that the dimension of the ImL_M equals the rank of M.

[4] Search for orthogonal bases of the solution spaces of the systems of linear equations in [1] and [2].

[5] Show that if M_1, M_2 are $n \times n$ matrices, dimension of $ImL_{M_1} \circ L_{M_2}$ =dimension of $ImL_{M_2} \circ L_{M_1}$. Here we assume that $F = \mathbb{C}$ although the conclusion holds for any field.

CHAPTER 6. QUADRATIC FORMS

As we are well aware, a scalar product on a finite dimensional vector space is determined by its associated symmetric matrix and vice versa. But we have another important notion, the so-called quadratic form which gives rise to the scalar product.

§ 6-1. Definitions

Suppose that we have a vector space V over a field F of characteristic $p \neq 2$; suppose further that V has a scalar product $<,>$, i.e., a symmetric bilinear form on V.

Definition(6.1.1). We define the quadratic form determined by $<,>$ to be the function $q : V \longrightarrow F$ via $q(v) =< v, v > \forall v \in V$.

Proposition(6.1.2). *We may recover any value $< v, w >$ from values of q $\forall v, w \in V$.*

Proof. We observe the following 2 polarization identities : $\frac{1}{2}\{q(v+w) - q(v) - q(w)\} =< v, w >$ and $\frac{1}{4}\{q(v+w) - q(v-w)\} =< v, w >$. Both of these are easily verified by way of the bilinearity of the scalar product. Either one is available for our purpose.

Example(6.1.3). In view of example 5.1.2, if a scalar product on a finite dimensional vector space V over F is given by a symmetric matrix $A = (f_{ij})_{n\times n}$, then for any column vector $X = {}_t(x_1, \cdots, x_n), q(X) = {}_tXAX = \sum_{i,j=1}^{n} f_{ij}x_ix_j$ is obtained. In particular, if $V = F^n$ with the usual dot product, then $q(X) = \sum_{i=1}^{n} x_i^2$ because the associated symmetric matrix is just $A = I_n$ in this case.

Example(6.1.4). Suppose that $V = \mathbb{R}^2$ with the associated matrix of a scalar product with respect to the standard basis. If $q(X) = 3x^2 + 4xy - 5y^2$ is given, then the associated matrix A of its symmetric bilinear form is calculated as follows : Let $A = \begin{pmatrix} f_{11} & f_{12} \\ f_{21} & f_{22} \end{pmatrix}$ with $f_{12} = f_{21}$ and

$$\begin{aligned} q(X) &= q(x, y) \\ &= (x, y)\begin{pmatrix} f_{11} & f_{12} \\ f_{21} & f_{22} \end{pmatrix}\begin{pmatrix} x \\ y \end{pmatrix} \\ &= f_{11}x^2 + 2f_{12}xy + f_{22}y^2. \end{aligned}$$

Comparing both sides of the equation $3x^2 + 4xy - 5y^2 = f_{11}x^2 + 2f_{12}xy + f_{22}y^2$, we have $A = \begin{pmatrix} 3 & 2 \\ 2 & -5 \end{pmatrix}$. Conversely if an associated matrix A of a symmetric bilinear form with respect to the standard basis is given like $A = \begin{pmatrix} 3 & 2 \\ 2 & -5 \end{pmatrix}$, then its quadratic form looks like $q(X) = 3x^2 + 4xy - 5y^2$.

Exercises of § 6-1

[1] Suppose that $V = \mathbb{R}^2$ with a scalar product. If $q(X) = 4x^2 + 3xy - 6y^2$ is given with respect to the standard basis of V, what is the associated matrix of the scalar product ?

[2] Suppose that $V = \mathbb{R}^3$ with a scalar product. If $q(X) = 3x^2+4xy+5yz+6xz+7y^2-8z^2$ is given for $X = {}_t(x, y, z)$, what is the associated matrix of this scalar product ?

[3] Suppose that q_1, q_2 are quadratic forms of symmetrix bilinear forms β_1, β_2 respectively. Verify that $q_1 + q_2$ is the quadratic form corresponding to the symmetric bilinear form $\beta_1 + \beta_2$.

[4] Let $q(X) = \sum_{i=1}^{n} x_i^2$ of a column vector $X = {}_t(x_1, \cdots, x_n)$ be a quadratic form. What is its associated scalar product ?

§ 6-2. Index of nullity

Let V be a finite dimensional vector space over F and let V have a scalar product $<,>$. We are anxious to know whether or not, there exists an orthogonal basis of V even if we can't find it out in reality.

Proposition(6.2.1). *For any V as above, we have its orthogonal basis.*

Proof. We shall proceed by induction on $dimV = n$. If $dimV = 1$, the assertion is very clear. Suppose that our conclusion holds for $dimV < n$. Now let $dimV = n$ and let $\{v_1, \cdots, v_n\}$ be a basis of V. If $< v, v >= 0 \forall v \in V$, then obviously $\{v_1, \cdots, v_n\}$ becomes an orthogonal basis of V. Otherwise $\exists v_1 \in V$ such that $< v_1, v_1 >\neq 0$. So there is a basis of V including v_1 like $\{v_1, v_2, \cdots, v_n\}$.

Let $v_i' := v_i - \frac{<v_i,v_1>}{<v_1,v_1>} v_1$ for $i = 2, \cdots, n$. Then $< v_i', v_1 >= 0, i = 2, \cdots, n$. Because $v_1, \cdots, v_n$ are linearly independent, so are $v_1, v_2', \cdots, v_n'$. If we let the span of $v_2', \cdots, v_n'$ be S, then evidently $S = (Fv_1)^\perp$, and so $dimS = n - 1 = dim(Fv_1)^\perp$.

Finally applying induction hypothesis to S, we have an orthogonal basis $\{w_2, \cdots, w_n\}$ of $(Fv_1)^\perp$. If we put $w_1 := v_1$, then $\{w_1, \cdots, w_n\}$ is an obvious orthogonal basis of V.

Definition(6.2.2). Suppose that V is a finite dimensional vector space over $\mathbb{R}$ with a scalar product. The dimension of the subspace $\{w \in V| < w, v >= 0 \; \forall v \in V\}$ is said to be *the index of nullity of the scalar product.*

Let $\widetilde{V}$ be the above subspace $\{w \in V| < w, v >= 0 \; \forall v \in V\}$. It is then evident that the index of nullity of the scalar product is zero if and only if the scalar product is non-degenerate. Such an index has a special meaning as the following proposition shows.

Proposition(6.2.3). *Let V and $\widetilde{V}$ be as above. Suppose that $\{\mathcal{O}_1, \cdots, \mathcal{O}_n\}$ is an orthogonal basis of V.*

We then have $dim\widetilde{V}$ =the cardinality of $\{i| < \mathcal{O}_i, \mathcal{O}_i >= 0\}$.

Proof. Let $\{\mathcal{O}_1, \cdots, \mathcal{O}_n\}$ rearranged so that $< \mathcal{O}_i, \mathcal{O}_i >\neq 0$ for $i \leq m$ and $< \mathcal{O}_j, \mathcal{O}_j >= 0$ for $j > m$. But then the orthogonality ensures that $\widetilde{V}$ contains $\mathcal{O}_{m+1}, \cdots, \mathcal{O}_n$. If an arbitrary vector w of $\widetilde{V}$ is expressed as $w = \sum_{i=1}^{n} r_i\mathcal{O}_i$ with $r_i \in \mathbb{R}$, then it gives $0 =< w, \mathcal{O}_i >=< \sum r_i\mathcal{O}_i, \mathcal{O}_i >= r_i < \mathcal{O}_i, \mathcal{O}_i >= 0$. So for $i \leq m, r_i = 0$ showing that $\{\mathcal{O}_{m+1}, \cdots, \mathcal{O}_n\}$ is a basis of $\widetilde{V}$ as required.

Exercises of § 6-2

[1] Show that proposition 6.2.3 also holds for any finite dimensional vector space with a scalar product over any field.

[2] What is the index of nullity for scalar products determined by the following symmetric matrices ?

$$\text{(i)} \begin{pmatrix} 1 & 3 \\ 3 & -2 \end{pmatrix} \qquad \text{(ii)} \begin{pmatrix} 4 & -2 \\ -2 & 3 \end{pmatrix}$$

§ 6-3. Sylvester theorem

In the preceding section, we have seen index of nullity may be computed by way of an orthogonal basis, i.e., we have noted some invariant of an orthogonal basis. We have another invariants of it in addition.

Theorem(6.3.1). *(**Sylvester theorem**) Suppose that a finite dimensional vector space V over $\mathbb{R}$ has a scalar product. For any orthogonal basis of $\{\mathcal{O}_1, \cdots, \mathcal{O}_n\}$ of V, the cardinality of $\{i | 0 >< \mathcal{O}_i, \mathcal{O}_i >\}$ is invariant.*

Proof. For any two orthogonal $\{\mathcal{O}_1, \cdots, \mathcal{O}_n\}$, $\{\mathcal{O}'_1, \cdots, \mathcal{O}'_n\}$, we rearrange them in such a way that

$$\begin{aligned} &< \mathcal{O}_i, \mathcal{O}_i >< 0 \quad \text{if} \quad 1 \leq i \leq \ell, \\ &< \mathcal{O}_i, \mathcal{O}_i >> 0 \quad \text{if} \quad \ell + 1 \leq i \leq m, \\ &< \mathcal{O}_i, \mathcal{O}_i >= 0 \quad \text{if} \quad m + 1 \leq i \leq n, \end{aligned}$$

and similarly

$$\begin{aligned} &< \mathcal{O}'_i, \mathcal{O}'_i >< 0 \quad \text{if} \quad 1 \leq i \leq \ell', \\ &< \mathcal{O}'_i, \mathcal{O}'_i >> 0 \quad \text{if} \quad \ell' + 1 \leq i \leq m', \\ &< \mathcal{O}'_i, \mathcal{O}'_i >= 0 \quad \text{if} \quad m' + 1 \leq i \leq n, \end{aligned}$$

where n is the dimension of V.

We consider first the set $\{\mathcal{O}_1, \cdots, \mathcal{O}_\ell\} \cup \{\mathcal{O}'_{\ell'+1}, \cdots, \mathcal{O}'_n\}$. For the purport of proving that this set is linearly independent, we assume that there is a relation $\sum_{i=1}^{\ell} r_i \mathcal{O}_i + \sum_{i=\ell'+1}^{n} r_i' \mathcal{O}_i' = 0$ with $r_i, r_i' \in \mathbb{R}$. So $\sum_{i=1}^{\ell} r_i \mathcal{O}_i = -\sum_{i=\ell'+1}^{n} r_i' \mathcal{O}_i'$ happens. But then it gives

$$< \sum_{i=1}^{\ell} r_i \mathcal{O}_i, \sum_{i=1}^{\ell} r_i \mathcal{O}_i > = < -\sum_{i=\ell'+1}^{n} r_i' \mathcal{O}_i', -\sum_{i=\ell'+1}^{n} r_i' \mathcal{O}_i' >,$$

from which it follows that

$$\sum_{i=1}^{\ell} r_i^2 < \mathcal{O}_i, \mathcal{O}_i > = \sum_{j=\ell'+1}^{n} (r_j')^2 < \mathcal{O}_j', \mathcal{O}_j' > .$$

The right hand side of this equation becomes ≥ 0, whereas the left hand side becomes ≤ 0. Hence we have $r_i = r_j' = 0\ \forall i = 1, \cdots, \ell,\ \forall j = \ell'+1, \cdots, n$, so that $\{\mathcal{O}_1, \cdots, \mathcal{O}_\ell, \mathcal{O}_{\ell'+1}', \cdots, \mathcal{O}_n'\}$ is a linearly independent set. Considering $dim V = n$, we have $\ell + (n - \ell') \leq n$. So $\ell \leq \ell'$ arises. Likewise if we think of the set $\{\mathcal{O}_1', \cdots, \mathcal{O}_{\ell'}', \mathcal{O}_{\ell+1}, \cdots, \mathcal{O}_n\}$, we get an inequality $\ell' \leq \ell$. Hence $\ell = \ell'$ is proven, as required.

Definition(6.3.2). The invariant of the cardinality of $\{i | < \mathcal{O}_i, \mathcal{O}_i > < 0\}$ makes us call the cardinality the *index of negativity* of the scalar product. Likewise $\{i | < \mathcal{O}_i, \mathcal{O}_i > > 0\}$ has the cardinality as an invariant which is called the *index of positivity* of the scalar product.

Remark(6.3.3). For a vector space V over $\mathbb{R}$ as above, we may call an orthogonal basis $\{\mathcal{O}_i, \cdots, \mathcal{O}_n\}$ othonormal in case $<$

$\mathcal{O}_i, \mathcal{O}_i >= 0$ or $< \mathcal{O}_i, \mathcal{O}_i >= \pm 1 \forall i = 1, \cdots, n$. For any orthogonal basis $\{\mathcal{O}_i, \cdots, \mathcal{O}_n\}$ of V, we put $\mathcal{O}'_i = \mathcal{O}_i$ if $< \mathcal{O}_i, \mathcal{O}_i >= 0$; we put

$$\mathcal{O}'_i = \frac{\mathcal{O}_i}{< \mathcal{O}_i, \mathcal{O}_i >^{\frac{1}{2}}} \quad \text{if} \quad < \mathcal{O}_i, \mathcal{O}_i >> 0;$$

we put

$$\mathcal{O}'_i = \frac{\mathcal{O}_i}{\{- < \mathcal{O}_i, \mathcal{O}_i >\}^{\frac{1}{2}}} \quad \text{if} \quad < \mathcal{O}_i, \mathcal{O}_i >< 0.$$

In this case we obtain an orthonormal basis $\{\mathcal{O}'_i, \cdots, \mathcal{O}'_n\}$ of V. From scratch, let $\{\mathcal{O}_i, \cdots, \mathcal{O}_n\}$ be an orthonormal basis of V. We then have that $\forall$ coordinate vector $\mathbb{X}$ of V with respect to the orthonormal basis of V, $q(\mathbb{X}) = -x_1^2 - \cdots - x_\ell^2 + x_{\ell+1}^2 + \cdots + x_m^2$, where $\mathbb{X} = {}_t(x_1, \cdots, x_n)$.

Exercises of § 6-3

[1] Let $V = \mathbb{R}^2$ with a scalar product determined by the matrix $\begin{pmatrix} -1 & 1 \\ 1 & -1 \end{pmatrix}$ with respect to the standard basis of V.

(i) Show that $\{{}_t(1,1), {}_t(1,0)\}$ is an orthogonal basis of V with respect to this form.

(ii) Also show whether or not this becomes an orthonormal basis of V.

[2] What is the index of negativity or the index of positivity or the index of nullity for each scalar product having its associated symmetric matrices on $\mathbb{R}^2$ as follows :

(i) $\begin{pmatrix} 2 & 3 \\ 3 & -1 \end{pmatrix}$ (ii) $\begin{pmatrix} 1 & -1 \\ -1 & 1 \end{pmatrix}$ (iii) $\begin{pmatrix} 3 & -5 \\ -5 & 4 \end{pmatrix}$

[3] Let $V = \mathbb{R}^n$ with its usual dot product. What is then the index of nullity or the index of positivity or the index of negativity of this symmetric bilinear form ?

CHAPTER 7. IMPORTANT OPERATORS

It seems very notable to find some important operators whose associated matrices may be diagonal with respect to some particular bases. Among these real symmetric matrices, hermitian matrices and unitary operators are noteworthy. We should understand that we need at least a nondegenerate scalar product for these operators in this chapter.

§ 7-1. Nondegenerate scalar product and symmetric operator

Suppose that a finite dimensional vector space V has a nondegenerate scalar product $<,>$. If there is another scalar product on V over F, we may ask whether or not it has something to do with $<,>$.

Proposition(7.1.1). *With assumptions as above, let β be any bilinear form on V. We then have unique operators $\mathcal{O}, \mathcal{P}$ of V satisfying that $\forall v, w \in V, \beta(v, w) =< \mathcal{O}v, w >=< v, \mathcal{P}w >$.*

Proof. By proposition 5.2.6, the functional $v \mapsto \beta(v, w)$ on V determines a unique vector $\widetilde{w}$ satisfying $L_{\widetilde{w}}(v) = \beta(v, w) \forall v, w \in$

V. So we have a linear map $\mathcal{P} : V \longrightarrow V$ via $\mathcal{P}(w) = \widetilde{w}$, so that $< v, \mathcal{P}w >=< v, \widetilde{w} >= L_{\widetilde{w}}(v) = \beta(v,w) \forall v, w \in V$. On the other hand the functional $w \mapsto \beta(v,w)$ on V could have yielded just as well an operator $\mathcal{O} : V \longrightarrow V$ via $\mathcal{O}(v) = \widetilde{v}$ satisfying $< \mathcal{O}v, w >=< \widetilde{v}, w >= \beta(v,w) \forall v, w \in V$. It is not difficult to show that $\mathcal{O}$ becomes linear.

Definition(7.1.2). In the preceding proposition the operator $\mathcal{P}$ is called *the transpose* of $\mathcal{O}$. If the transpose of $\mathcal{O}$ is $\mathcal{O}$ itself, $\mathcal{O}$ is called *symmetric* with respect to the fixed $<,>$. We shall write the transpose of $\mathcal{O}$ as ${}_t\mathcal{O}$.

Proposition(7.1.3). *Suppose that a finite dimensional vector space V over F has a nondegenerate symmetric bilinear form, i.e., a nondegenerate scalar product; then $\forall f \in F$, $\forall$ operators $\mathcal{O}, \mathcal{P}$, we have*

$$\begin{aligned} (i) &\quad {}_{tt}\mathcal{O} = \mathcal{O} \\ (ii) &\quad {}_t(\mathcal{O} + \mathcal{P}) = {}_t\mathcal{O} + {}_t\mathcal{P} \\ (iii) &\quad {}_t(f\mathcal{O}) = f\,{}_t\mathcal{O} \\ (iv) &\quad {}_t(\mathcal{O}\mathcal{P}) = {}_t\mathcal{P}\,{}_t\mathcal{O}. \end{aligned}$$

Proof. We verify only (i) the others being just as simple to do. Now $\forall v, w \in V$, we have

$$\begin{aligned} &< v, {}_{tt}\mathcal{O}w >=< {}_t\mathcal{O}v, w > \\ &= < w, {}_t\mathcal{O}v >=< \mathcal{O}w, v > \\ &= < v, \mathcal{O}w > . \end{aligned}$$

Hence it follows from nondegeneracy that ${}_{tt}\mathcal{O} = \mathcal{O}$.

Example(7.1.4). In order to justify the name of transpose, we consider $V = {}_t(F^m)$ with the usual dot product $<,>$. For any column vectors $\mathbb{X}, \widetilde{\mathbb{X}} \in {}_t(F^m)$ and for any operator $\mathcal{O}$, we have $< \mathcal{O}\mathbb{X}, \widetilde{\mathbb{X}} >= {}_t(\mathcal{O}\mathbb{X})\widetilde{\mathbb{X}} = {}_t\mathbb{X}_t\mathcal{O}\widetilde{\mathbb{X}}$. Here we may view $\mathcal{O}$ as a matrix with respect to the standard basis. So the rightmost side of the last equation is easily understood.

On the other hand $< \mathbb{X}, {}_t\mathcal{O}\widetilde{\mathbb{X}} >= {}_t\mathbb{X}_t\mathcal{O}\widetilde{\mathbb{X}} =< \mathcal{O}\mathbb{X}, \widetilde{\mathbb{X}} >= {}_t\mathbb{X}_t\mathcal{O}\widetilde{\mathbb{X}}$. It follows from this that the transpose of an operator is expressed as the transpose of its associated matrix, which justifies the name.

Remark(7.1.5). Let V be a vector space as in proposition 7.1.1 with a scalar product $<,>$; further let $\mathcal{O}$ be an invertible symmetric operator.

If we define $\forall v, w \in V, < \mathcal{O}v, w >=:<< v, w >>$, then $<<, >>$ becomes another nondegenerate scalar product on V and vice versa in view of proposition 7.1.1. So we may have quadratic forms $q(v) =< v, v >$ and $q_{\mathcal{O}}(v) =< \mathcal{O}v, v >=<< v, v >>$ respectively, which forces us to call the latter *quadratic form determined by* $\mathcal{O}$.

Exercises of § 7-1

[1] What is the quadratic form determined by the matrix $\begin{pmatrix} 4 & 3 \\ 3 & -2 \end{pmatrix}$ acting on $V = \mathbb{R}^2$?

[2] Prove proposition 7.1.3 completely.

[3] If a matrix M equals $-_tM$, we call it skew-symmetric. For any square matrix A, show that $A - {}_tA$ becomes skew-symmetric and that A may be written as a sum of a skew-symmetric matrix and a symmetric one uniquely determined by using $A - {}_tA$.

[4] For any $m \times m$ square matrix $M = (f_{ij})_{m\times m}$, we define the trace of M to be the sum of diagonal entries, i.e., $tr(M) = \sum_{i=1}^{m} f_{ii}$:

(i) show that $tr(M_1M_2) = tr(M_2M_1)$ $\forall$square matrices M_1, M_2.

(ii) show that the trace of any operator $\mathcal{O}$ may be defined by its associated matrix and that $tr(\mathcal{O}) = tr({}_t\mathcal{O})$.

[5] Let M be a $m \times m$ matrix and A be an invertible matrix. Verify that $tr(M) = tr(A^{-1}MA)$ and that $tr\mathcal{O}$ is well-defined for any operator $\mathcal{O}$.

§ 7-2. Positive definite hermitian form and Hermitian operator

We restate the latter contents of § 5-1 because of their importance. Suppose that V is a vector space over the field $\mathbb{C}$ of complex numbers. It is meaningless to think of a scalar product on V considering the complex plane. Instead we think of some variant of scalar product explained in the following as in § 5-1.

Definition(7.2.1). By a hermitian 2product (or form) on V over $\mathbb{C}$, we mean a function $<,>: V \times V \longrightarrow \mathbb{C}$ such that the following properties are satisfied :

(i) If we denote the complex conjugate by the notation$-$(bar), then $\forall v, w \in V <w, v>=\overline{<v, w>}$.

(ii) $\forall v_1, v_2, w \in V, <v_1 + v_2, w>=<v_1, w> + <v_2, w>$.

(iii) $\forall c \in \mathbb{C}, \forall v, w \in V, <cv, w>= c <v, w>$ but $<v, cw>= \bar{c} <v, w>$.

Furthermore if $\forall v (\neq 0) \in V$, $<v, v>> 0$ holds, we say that the hermitian product is positive definite. As with chapter 5, we may also define similar terminologies to nondegeneracy, perpendicularity, orthogonality, orthogonal basis, orthonormality, norm etc. in this situation. A typical and easy example of a hermitian product is at hand as shown in example 5.1.17.

Example(7.2.2). If $V = \mathbb{C}^m$, then $\forall$ vectors $v = (v_1, \cdots, v_m)$, $w = (w_1, \cdots, w_m) \in \mathbb{C}^m$, we define $<v, w>:= \sum_{i=1}^m v_i \bar{w}_i$, which is easily to be seen a hermitian product. In addition it is positive define because $\forall v = (v_1, \cdots, v_m)(\neq 0) \in V$, $\|v\|^2 :=< v, v >= \sum_{i=1}^m v_i\bar{v_i} = \sum_{i=1}^m |v_i|^2 > 0$ holds.

Analogues of previous facts regarding positive definite scalar product may be set up when given a positive definite hermitian product. Explicit proofs of the following statements are left to the readers.

Proposition(7.2.3). *(**Gram-Schmidt process**) Suppose that V is a finite dimensional vector space over $\mathbb{C}$ with a positive definite hermitian product and that S is a proper subspace of V with an orthogonal basis of S; then we may extend the orthogonal basis of S to that of V.*

Proof. Refer to theorem 5.1.13. Note that here $S = 0$ may happen.

Proposition(7.2.4). *Suppose that V is a finite dimensional vector space with a positive definite hermitian product and that S is a subspace of V; then we have $V = S \oplus S^\perp$.*

Proof. Refer to corollary 5.1.15.

Example(7.2.5). Let V be a finite dimensional vector space over $\mathbb{C}$ with a positive definite hermitian product. If we define $\|v\| :=< v, v >^{\frac{1}{2}} \forall v \in V$, then we may show $| < v, w > | \leq \|v\|\|w\| \forall v, w \in V$. For its proof, let $\{v_1, \cdots, v_\ell\}$ be an orthonormal basis of V. Then we may view V as $\mathbb{C}^\ell$ with the usual hermitian product as in example 7.2.2. Let A be any vector and let

$c :=< A, E >$ for a unit vector E. Then $A - cE$ is perpendicular to cE. So we have $\|A\|^2 = \|A-cE\|^2 + \|cE\|^2 = \|A-cE\|^2 + |c|^2$, so that $|c| \leq \|A\|$. Next we put $E := \frac{B}{\|B\|}$ and get $\frac{|<A,B>|}{\|B\|} \leq \|A\|$. Hence $| < A, B > | \leq \|A\|\|B\|$ holds.

Proposition(7.2.6). *If V is a finite dimensional vector space over $\mathbb{C}$ with a positive definite hermitian form $<,>$ and if ρ is a functional on V, then we may find $\widetilde{w}$ uniquely satisfying $\rho(v) =< v, \widetilde{w} > \forall v \in V$. In particular we have an operator $\mathcal{P}$ of V such that $< \mathcal{O}v, w >=< v, \mathcal{P}w > \forall v, w \in V$ for any operator $\mathcal{O}$ of V.*

Proof. Just refer to the proof shown in the real case.

Definition(7.2.7). In the preceding proposition the operator $\mathcal{P}$ is called *the adjoint* of $\mathcal{O}$ and denoted by $_*\mathcal{O}$. If $_*\mathcal{O} = \mathcal{O}$, $\mathcal{O}$ is called *hermitian* or *self-adjoint.*

Example(7.2.8). If we think of $V = \mathbb{C}^\ell$ with the usual hermitian form, then for any complex matrix $M, {_*M} = {_t\overline{M}}$ happens because $< \mathbb{X}, \mathbb{Y} >= {_t\mathbb{X}}\overline{\mathbb{Y}}$ for any column vectors $\mathbb{X}, \mathbb{Y}$.

Proposition(7.2.9). *Suppose that a finite dimensional vector space V over $\mathbb{C}$ has a positive definite hermitian form ; then $\forall c \in \mathbb{C}$, $\forall$ operators $\mathcal{O}, \mathcal{P}$ of V*

(i) $_{**}\mathcal{O} = \mathcal{O}$

(ii) $_*(c\mathcal{O}) = \bar{c}_*\mathcal{O}$

(iii) $_*(\mathcal{O}\mathcal{P}) = {_*\mathcal{P}}_*\mathcal{O}$

(iv) $_*(\mathcal{O} + \mathcal{P}) = {_*\mathcal{O}} + {_*\mathcal{P}}$.

Proof. Left to the readers.

Proposition(7.2.10). *With assumptions as above let* $<<,>>$ *be another hermitian form on* V; *then we have a unique hermitian operator* $\mathcal{O}$ *of* V *satisfying* $<< v, w >>=< \mathcal{O}v, w > \forall v, w \in V$.

Proof. All in all the proof comes from proposition 7.2.6.

Proposition(7.2.11). *With assumptions as before if* $< \mathcal{O}v, v > = 0 \ \forall v \in V$, *then* $\mathcal{O} = 0$ *(trivial).*

Proof. Since

$$< \mathcal{O}(v + w), v + w > - < \mathcal{O}(v - w), v - w >$$
$$= 2 < \mathcal{O}v, w > +2 < \mathcal{O}w, v > \forall v, w \in V,$$

we get $0 =< \mathcal{O}v, w > + < \mathcal{O}w, v >$. Substituting iv for v yield $0 =< \mathcal{O}v, w > - < \mathcal{O}w, v >$, so that we have $< \mathcal{O}v, w >= 0$, i.e., $\mathcal{O} = 0$ comparing the last two equations.

Proposition(7.2.12). *Under preceding assumption an operator* $\mathcal{O}$ *is hermitian if and only if* $\forall v \in V, < \mathcal{O}v, v >$ *is real.*

Proof. Suppose $< \mathcal{O}v, v >$ is real ; then $< \mathcal{O}v, v >=< v, \mathcal{O}v >= < {}_*\mathcal{O}v, v >$, so that $< ({}_*\mathcal{O} - \mathcal{O})v, v >= 0 \forall v \in V$. It follows that ${}_*\mathcal{O} - \mathcal{O} = 0$ from proposition 7.2.11. Hence $\mathcal{O}$ is hermitian. For the converse look at $< \mathcal{O}v, v >=< v, {}_*\mathcal{O}v >=< v, \mathcal{O}v >= < \overline{\mathcal{O}v, v} >$, which means $< \mathcal{O}v, v >$ is real $\forall v \in V$.

Exercises of § 7-2

[1] Let V be a finite dimensional vector space over $\mathbb{C}$. If v, w are perpendicular, show that $\|v+w\|^2 = \|v\|^2 + \|w\|^2$, where we assume V has a positive definite hermitian product.

[2] Show that any positive definite hermitian product is also nondegenerate.

[3] Let V be $\mathbb{C}^4$ with the usual hermitian product. What is an orthonormal basis for the subspace of V generated by the elements $(i, 2, -i, 1)$ and $(1, i, 2, -i)$?

[4] Prove all propositions of this section if necessary.

§ 7-3. Real unitary operator

Suppose that a positive definite scalar product is given to a finite dimensional vector space over $\mathbb{R}$.

Definition(7.3.1). If an operator $\mathcal{O} : V \longrightarrow V$ satisfies $< v, w >=< \mathcal{O}v, \mathcal{O}w > \ \forall v, w \in V$, then $\mathcal{O}$ is said to be a *unitary* or an *orthogonal map.*

Some other equivalent notions are at hand.

Proposition(7.3.2). *Under assumptions as above the following are equivalent :*

(i) $\forall v \in V, \ \|\mathcal{O}v\| = \|v\|$,

(ii) $\forall$ *unit vector* $v \in V, \ \|\mathcal{O}v\| = \|v\| = 1$,

(iii) $\mathcal{O}$ *is a unitary operator.*

Proof. (i) $\Longrightarrow$ (ii) is easy. For (ii) $\Longrightarrow$ (i), we note $\forall v(\neq 0) \in V$, $\frac{v}{\|v\|}$ is a unit vector. So $\|\mathcal{O}\frac{v}{\|v\|}\| = \|\frac{v}{\|v\|}\| = 1$ holds. But then $\|\mathcal{O}v\| = \|v\|$ happens.

Now for (i) $\Longrightarrow$ (iii), we note $< \mathcal{O}v, \mathcal{O}v >=< v, v > \forall v \in V$. So from $< \mathcal{O}(v+w), \mathcal{O}(v+w) > - < \mathcal{O}(v-w), \mathcal{O}(v-w) >= 4 < \mathcal{O}v, \mathcal{O}w >$, it follows that $< v+w, v+w > - < v-w, v-w >= 4 < v, w >= 4 < \mathcal{O}v, \mathcal{O}w >$, and hence $< v, w >=< \mathcal{O}v, \mathcal{O}w >$. Since $< \mathcal{O}v, \mathcal{O}v >= \|\mathcal{O}v\|^2$, (iii) $\Longrightarrow$ (i) follows at once.

Proposition(7.3.3). *With assumptions as before* $\mathcal{O} : V \longrightarrow V$ *is unitary if and only if* ${}_t\mathcal{O}\mathcal{O} = id$ *holds.*

Proof. $\forall v, w \in V, < \mathcal{O}v, \mathcal{O}w >=< v, w >=< v, {}_t\mathcal{O}\mathcal{O}w > \Longleftrightarrow {}_t\mathcal{O}\mathcal{O} = id$ by virtue of proposition 7.1.1.

Definition(7.3.4). In view of this proposition we define a real $\ell \times \ell$ matrix M to be unitary if and only if ${}_tMM = I_\ell$.

Example(7.3.5). We pause for a moment to consider all unitary operators of $\mathbb{R}^2$. $\forall v, w \in \mathbb{R}^2$ we see at once $< \mathcal{O}v, \mathcal{O}w >= \|\mathcal{O}v\|\|\mathcal{O}w\|cos\theta = \|v\|\|w\|cos\theta =< v, w >$ with θ an angle between v and w.

So $\mathcal{O}$ is nothing else but a rigid motion as a linear map. Hence $\mathcal{O}$ is a rotation or a reflection or a composite of them. It is not difficult to see that a rotation is represented by the matrix $\begin{pmatrix} cos\theta & sin\theta \\ sin\theta & +cos\theta \end{pmatrix}$ and a reflection is represented by the matrix $\begin{pmatrix} cos\theta & +sin\theta \\ sin\theta & cos\theta \end{pmatrix}$ in $\mathbb{R}^2$.

Exercises of § 7-3

[1] Show that if an operator is real unitary, then it is invertible.

[2] Verify that composites of real unitary maps are also unitary.

[3] If $\mathcal{O}$ is a diagonal real unitary matrix, show then that its diagonal entries are ± 1.

[4] If $\mathcal{O}$ is unitary, show then that any orthonormal basis is invariant under $\mathcal{O}^2 \circ {}_t\mathcal{O}^2$, i.e., $\mathcal{O}^2 \circ {}_t\mathcal{O}^2 \mathcal{B} \subset \mathcal{B}$ for an orthonormal basis $\mathcal{B}$ of V.

§ 7-4. Complex unitary operator

Analogues of the preceding facts are plausible in the case of V being a finite dimensional vector space over $\mathbb{C}$ with a positive definite hermitian product.

Definition(7.4.1). With V as above, if an operator $\mathcal{O} : V \longrightarrow V$ satisfies $< v, w >=< \mathcal{O}v, \mathcal{O}w > \ \forall v, w \in V$, then $\mathcal{O}$ is called a *complex unitary operator.*

Similarly to proposition 7.3.2, we obtain

Proposition(7.4.2). *Under assumptions as above we have the following equivalent statements :*

(i) $\forall v \in V, \ \|\mathcal{O}v\| = \|v\|$.

(ii) $\forall$ *unit vector* $v \in V, \ \|\mathcal{O}v\| = \|v\| = 1$.

(iii) $\mathcal{O}$ *is complex unitary.*

Proof. Just imitate word for word the proofs of proposition 7.3.2.

Proposition(7.4.3). *Under assumptions as above* $\mathcal{O} : V \longrightarrow V$ *is complex unitary if and only if* ${}_*\mathcal{O}\mathcal{O} = id$.

Proof. The proof is as simple as that of proposition 7.3.3.

Example(7.4.4). Let $V = \mathbb{C}^\ell$ with the usual hermitian form; then we have seen in example 7.2.8 that for any complex matrix M, ${}_*M = {}_t\bar{M}$. So we could also have defined a complex matrix M is unitary if and only if ${}_t\bar{M}M = I_\ell$ in view of proposition 7.4.3.

Remark(7.4.5). In the next chapter we shall see that symmetric operators, hermitian operators, and complex unitary operators may have a certain basis with respect to which their associated matrices become diagonal.

Exercises of § 7-4

[1] Show that a composite of any two complex unitary operators also becomes unitary.

[2] Show that for any complex unitary $\mathcal{O}$ and for any orthonormal basis $\mathcal{B}$, $\mathcal{B}$ is invariant under $\mathcal{O}^m \cdot {}_*\mathcal{O}^m$ with $m \in \mathbb{Z}^{>0} = \mathbb{N}$ the natural numbers.

[3] Verify that if $\mathcal{O}$ is a diagonal complex unitary matrix, then its diagonal entries are of the form $e^{i\theta}$ for $0 \leq \theta < 2\pi$.

[4] Show that for any orthonormal basis $\mathcal{B}$ and for any complex unitary $\mathcal{O}$, $\mathcal{O}\mathcal{B}$ is also an orthonormal basis.

CHAPTER 8. DIAGONALIZATION OF CERTAIN MATRICES

Certain operators may be diagonalized by way of certain bases. If the vector space under consideration has either a positive definite scalar product or a positive definite hermitian form, then it gives rise to an orthonormal basis which affects diagonalization of certain operators.

§ 8-1. Eigenvalues and Eigenvectors

Henceforth we shall meet with certain important concepts frequently which shall be introduced in what follows.

Definition(8.1.1). Let $\mathcal{O}$ be an operator of a vector space V over a field F. A nonzero vector $v \in V$ is called *an eigenvector* of $\mathcal{O}$ if $\exists \varepsilon \in F$ such that $\mathcal{O}v = \varepsilon v$ and ε is called *an eigenvalue* of $\mathcal{O}$ relative to v in this case. $V_\varepsilon := \{v \in V | \mathcal{O}v = \varepsilon v\}$ is called the *eigenspace* of $\mathcal{O}$ relative to ε.

Proposition(8.1.2). *The eigenspace V_ε in the preceding definition is a subspace of V.*

Proof. $\forall v, w \in V, \forall f \in F$, we have $\mathcal{O}(v+w) = \mathcal{O}v + \mathcal{O}w = \varepsilon v + \varepsilon w = \varepsilon(v+w)$ and $\mathcal{O}(fv) = f\mathcal{O}v = f\varepsilon v = \varepsilon(fv)$. $\therefore$ V_ε is a subspace.

Proposition(8.1.3). *Suppose that certain eigenvectors of $\mathcal{O}$ have distinct eigenvalues; then they are linearly independent vectors.*

Proof. Let eigenvectors $\{v_1, \cdots, v_\ell\}$ have distinct eigenvalues $\{\varepsilon_1, \cdots, \varepsilon_\ell\}$ of $\mathcal{O}$. We shall proceed by induction on ℓ. Nothing to show for $\ell = 1$. After induction hypothesis, we assume that for some $f_i \in F$, there is a relation $\sum_{i=1}^{\ell} f_i v_i = 0$. So $\sum_{i=1}^{\ell} \varepsilon_1 f_i v_i = 0$ holds. Also $0 = \mathcal{O}(\sum_{i=1}^{\ell} f_i v_i) = \sum_{i=1}^{\ell} f_i \varepsilon_i v_i$ holds.

Subtracting the last two equations yields $\sum_{i=2}^{\ell}(\varepsilon_1 - \varepsilon_i) f_i v_i = 0$. But then $\varepsilon_1 - \varepsilon_i \neq 0$ for $i = 2, \cdots, \ell$, so that we have $f_2 = \cdots = f_\ell = 0$ by induction hypothesis. Finally $f_1 = 0$ is obtained from the relation.

Proposition(8.1.4). *Suppose that $\mathcal{O} : V \longrightarrow V$ is an operator of a finite dimensional vector space over F. In this case $\mathcal{O}$ has an eigenvalue ε if and only if $\mathcal{O} - \varepsilon I$ is not an invertible operator of V.*

Proof. (necessity) Let v be a nonzero eigenvector with its eigenvalue ε of $\mathcal{O}$. So $\mathcal{O}v = \varepsilon v$ yields $(\mathcal{O} - \varepsilon I)v = 0$, which shows $\mathcal{O} - \varepsilon I$ is not invertible.

Conversely for the (sufficiency) we suppose $\mathcal{O} - \varepsilon I$ is not invertible. So it is easy to see that $Ker(\mathcal{O} - \varepsilon I) \neq \{0\}$. Let v

be any nonzero vector in $Ker(\mathcal{O} - \varepsilon I)$; then $(\mathcal{O} - \varepsilon I)v = 0$, i.e., $\mathcal{O}v = \varepsilon v$. So ε is an eigenvalue of $\mathcal{O}$.

Example(8.1.5). Let $\mathcal{B}$ be the standard basis of ${}_t(F^\ell)$ consisting of standard unit vectors ${}_tU_i$. Let $\mathcal{O}$ be an operator of ${}_t(F^\ell)$ and let $A_{\mathcal{B}}^{\mathcal{B}}(\mathcal{O}) = diag(d_1, \cdots, d_\ell)$. In this case $(\mathcal{O} - d_iI)_tU_i = d_{it}U_i - d_{it}U_i = 0$ because $\mathcal{O}$ has eigenvalues d_i relative to ${}_tU_i$. The above proposition generalizes such a simple fact that $(\mathcal{O} - d_iI)$ is not invertible.

Proposition(8.1.6). *Suppose that M is an $\ell \times \ell$ matrix over F and that M viewed as an operator of ${}_t(F^\ell)$ has ℓ distinct eigenvalules. We then have a basis $\mathcal{B}$ consisting of eigenvectors such that M may be diagonalized by using $\mathcal{B}$.*

Proof. By virtue of proposition 8.1.3, the set $\widetilde{\mathcal{B}}$ consisting of eigenvectors $v_1, \cdots, v_\ell$ corresponding to distinct eigenvalues $\varepsilon_1, \cdots, \varepsilon_\ell$ becomes a basis of ${}_t(F^\ell)$.

Let $\mathcal{O} : {}_tF^\ell \longrightarrow {}_tF^\ell$ be an operator such that for the standard basis $\mathcal{B}$ of ${}_tF^\ell, A_{\mathcal{B}}^{\mathcal{B}}(\mathcal{O}) = M$ holds. By virtue of proposition 4.3.8, we then have $A_{\widetilde{\mathcal{B}}}^{\mathcal{B}}(id) \circ A_{\mathcal{B}}^{\mathcal{B}}(\mathcal{O}) \circ A_{\mathcal{B}}^{\widetilde{\mathcal{B}}}(id) = A_{\widetilde{\mathcal{B}}}^{\widetilde{\mathcal{B}}}(\mathcal{O})$. But then $A_{\widetilde{\mathcal{B}}}^{\mathcal{B}}(id) \circ A_{\mathcal{B}}^{\widetilde{\mathcal{B}}}(id) = I_\ell$ and $A_{\widetilde{\mathcal{B}}}^{\widetilde{\mathcal{B}}}(\mathcal{O}) = diag(\varepsilon_1, \cdots, \varepsilon_\ell)$ ensures the assertion.

Example(8.1.7). We consider $M = \begin{pmatrix} 0 & 1 \\ -1 & 0 \end{pmatrix}$ over $\mathbb{R}$. Suppose that $M\begin{pmatrix} x \\ y \end{pmatrix} = \varepsilon \begin{pmatrix} x \\ y \end{pmatrix}$ for some $\varepsilon \in \mathbb{R}$. We then have $y = \varepsilon x$ and $-x = \varepsilon y$, so that $(\varepsilon^2 + 1)x = 0$. Since $\varepsilon^2 + 1 \neq 0$, there is no real eigenvalue. Of course M has a complex eigenvalues $\pm i$. So the field F is important here.

Example(8.1.8). Let $M = \begin{pmatrix} 1 & f \\ 0 & 1 \end{pmatrix}$ with $f \neq 0$ in any field F. Note that M is viewed as an operator via $(x, y) \overset{\theta}{\mapsto} (x + fy, y)$. If there is an eigenvalue ε relative to an eigenvector (x, y), then $x + fy = \varepsilon x$ and $y = \varepsilon y$.

We conclude from this $\varepsilon = 1$ is the only eigenvalue of $\mathcal{O}$ relative to the only eigenvectors $F(1, 0)$. But since $(1, 0)$ cannot span the whole space $V = F^2$, we may say that M cannot be similar to a diagonal matrix by virtue of proposition 8.1.6.

Definition(8.1.9). For any $\ell \times \ell$ matrix M over F. The *characteristic polynomial* $P_M(\mathbb{X})$ of $M = (f_{ij})$ is defined to be the determinant $D(\mathbb{X}I_\ell - M) = P_M(\mathbb{X})$ or $D(M - \mathbb{X}I_\ell) = P_M(\mathbb{X})$.

Proposition(8.1.10). *For $M = (f_{ij})_{\ell \times \ell}$ as above, r is a root of $P_M(\mathbb{X})$ if and only if r is an eigenvalue of M.*

Proof. (**Necessity**) Since $D(rI_\ell - M) = 0$, we see $rI_\ell - M : {}_tF^\ell \longrightarrow {}_tF^\ell$ viewed as an operator is not invertible by virtue of remark 3.3.3. So there is a nonzero $v \in {}_tF^\ell$ such that $(rI_\ell - M)v = 0$. (Otherwise $Ker(rI_\ell - M) = 0$ implies $rI_\ell - M$ is injective.) Namely $Mv = rv$ holds.

(**Sufficiency**) Because M has an eigenvalue r, $M - rI_\ell$ is not invertible in view of proposition 8.1.4. But then by remark 3.3.3, $D(rI_\ell - M) = 0$.

We shall use this proposition in the next section. Since similar matrices have the same characteristic polynomial, we could also have defined the characteristic polynomial $P_{\mathcal{O}}(\mathbb{X})$ of an operator $\mathcal{O}$ of V in case V is finite dimensional over F.

Exercises of § 8-1

[1] What are the eigenvalues and eigenvectors of the following matrices over $\mathbb{R}$ and $\mathbb{C}$ respectively ? :

$$\text{(i)} \begin{pmatrix} 3 & -2 \\ 2 & -3 \end{pmatrix} \quad \text{(ii)} \begin{pmatrix} -1 & 2 \\ -2 & 2 \end{pmatrix} \quad \text{(iii)} \begin{pmatrix} 2 & -4 \\ 5 & 3 \end{pmatrix}.$$

[2] Show that an $\ell \times \ell$ matrix M whose rows have the sum equal to n for each row of M has an eigenvalue n $\forall n \in \mathbb{N}$.

[3] Find the eigenvalues and eigenvectors of the matrix $\begin{pmatrix} 1 & 2 \\ -1 & 1 \end{pmatrix}$ over $\mathbb{C}$.

[4] Verify that $\begin{pmatrix} cos\theta & -sin\theta \\ sin\theta & cos\theta \end{pmatrix}$ has no eigenvector in $\mathbb{R}^2$ if θ is no integral multiple of π.

[5] If V is the $\mathbb{R}$-vector space generated by all infinitely differentiable functions, show that $e^{\alpha x}$ for $\alpha \in \mathbb{R}$ is an eigenvector of $\frac{d}{dx}$.

§ 8-2. Spectral basis theorem

Let V be a nonzero finite dimensional vector space over $\mathbb{R}$ with a positive definite scalar product. In this section we study real symmetric operator by and large.

Proposition(8.2.1). *Suppose that v is an eigenvector of a real symmetric operator $\mathcal{O}$ of V; then for any vector w orthogonal to v, v is also orthogonal to $\mathcal{O}w$. If a subspace S satisfies $\mathcal{O}(S) \subset S$, then $\mathcal{O}(S^{\perp}) \subset S^{\perp}$ also holds.*

Proof. Let $\mathcal{O}v = \varepsilon v$ for some $\varepsilon \in \mathbb{R}$. Because $\mathcal{O}$ is symmetric, we obtain $0 = \varepsilon < v, w >=< \varepsilon v, w >=< \mathcal{O}v, w >=< v, \mathcal{O}w >$. The latter assertion is trivial.

Proposition(8.2.2). *Suppose that an $\ell \times \ell$ real symmetric matrix M has an eigenvalue $\varepsilon \in \mathbb{C}$; then ε is real, i.e. $\varepsilon \in \mathbb{R}$.*

Proof. If $ME = \varepsilon E$ for some $E \in {}_t\mathbb{C}^{\ell}$, then for the usual hermitian form $<,>$ in ${}_t\mathbb{C}^{\ell}$, we have $< ME, E >=< \varepsilon E, E >= \varepsilon < E, E >$, while $< ME, E >=< E, {}_tME >=< E, ME >=< \overline{ME, E} >$ holds. But then $\varepsilon = \bar{\varepsilon}$, so that ε is real.

Proposition(8.2.3). *(***Spectral basis theorem***)*

For any real symmetric operator $\mathcal{O}$, we have an orthogonal basis of V consisting of eigenvectors of $\mathcal{O}$, and so an orthonormal basis of V exists, where V is a finite dimensional vector space over $\mathbb{R}$ with a positive definite scalar product.

Proof. After choosing any orthonormal basis of V, we may view $\mathcal{O}$ as a matrix operator $\mathcal{O} : {}_t\mathbb{R}^\ell \longrightarrow {}_t\mathbb{R}^\ell$ with $dimV = \ell$. By proposition 8.1.10, and proposition 8.2.2, we have an eigenvector e_1 of $\mathcal{O}$ in V. So if we put $S := \mathbb{R}e_1$, then $S \oplus S^\perp = V$ holds according to proposition 5.2.8.

So if we proceed inductively on ℓ in view of proposition 8.2.1, then the induction hypothesis assures the existence of an orthogonal basis $\{e_2, \cdots, e_\ell\}$ of $S^\perp$ consisting of eigenvectors of $S^\perp$. Hence $\{e_1, e_2, \cdots, e_\ell\}$ is an orthogonal basis of V. Evidently $\{\frac{e_i}{||e_i||} | i = 1, \cdots, \ell\}$ is an orthonormal basis.

Proposition(8.2.4). *Suppose that M is an $\ell \times \ell$ real symmetric matrix; we then have a real unitary matrix u satisfying $u^{-1}Mu = {}_tuMu$ is a diagonal one.*

Proof. We may view M as an associated matrix with an operator $\mathcal{O} : {}_t\mathbb{R}^\ell \longrightarrow {}_t\mathbb{R}^\ell$. By the spectral basis theorem we have an orthonormal basis (of ${}_t\mathbb{R}^\ell$) $\widetilde{\mathcal{B}}$ consisting of eigenvectors of ${}_t\mathbb{R}^\ell$. Let $\mathcal{B}$ be the basis consisting of standard unit vectors of ${}_t\mathbb{R}^\ell$ and let $\widetilde{\mathcal{B}} = \{e_1, \cdots, e_\ell\}$. We then have $A^{\mathcal{B}}_{\widetilde{\mathcal{B}}}(id) \circ A^{\mathcal{B}}_{\mathcal{B}}(\mathcal{O}) \circ A^{\widetilde{\mathcal{B}}}_{\mathcal{B}}(id) = A^{\widetilde{\mathcal{B}}}_{\widetilde{\mathcal{B}}}(\mathcal{O})$ which is a diagonal matrix. We put $u = A^{\widetilde{\mathcal{B}}}_{\mathcal{B}}(id)$ and contend that it is unitary.

In fact $id(e_i) = e_i = \sum_{k=1}^{\ell} r_{ki}U_k$ for $i = 1, \cdots, \ell$ and for some $r_{ki} \in \mathbb{R}$. Since $< e_i, e_j >= \sum_{k=1}^{\ell} r_{ki}r_{kj} = 0$ if $i \neq j$ and $< e_i, e_i >= \sum_{k=1}^{\ell} r_{ki}^2 = 1$, we have ${}_tA^{\widetilde{\mathcal{B}}}_{\mathcal{B}}(id) \circ A^{\widetilde{\mathcal{B}}}_{\mathcal{B}}(id) = {}_t(r_{ki})(r_{ki}) = I_\ell$. Hence ${}_tA^{\widetilde{\mathcal{B}}}_{\mathcal{B}}(id) = A^{\widetilde{\mathcal{B}}}_{\mathcal{B}}(id)^{-1}$ showing that $A^{\widetilde{\mathcal{B}}}_{\mathcal{B}}(id)$ is real unitary. Now putting $u := A^{\widetilde{\mathcal{B}}}_{\mathcal{B}}(id)$ finishes the proof.

Definition(8.2.5). If an operator $\mathcal{O}$ satisfies $\mathcal{O}(S) \subset S$ for a subspace of V as in proposition 8.2.1, we call S *$\mathcal{O}$-invariant.*

Exercises of § 8-2

[1] For any $\ell \times \ell$ real symmetric matrix M, verify that an orthonormal basis of $\mathbb{R}^\ell$ consisting of eigenvectors of M exists.

[2] If u is real unitary, show that u^{-1} is so.

[3] If $M = \begin{pmatrix} 2 & 1 \\ 1 & 3 \end{pmatrix}$, what is an orthogonal basis of $\mathbb{R}^2$ consisting of eigenvectors of M ?

§ 8-3. Hermitian operator

Suppose that we are given a finite dimensional vector space V over $\mathbb{C}$ with a positive definite hermitian form. Hermitian operator takes after real symmetric operator in some aspects.

Proposition(8.3.1). *For V as above, any eigenvalue ε of a hermitian operator $\mathcal{O} : V \longrightarrow V$ is a real number.*

Proof. If $\mathcal{O}v = \varepsilon v$ for some eigenvector $v \in V$, we then have $< \mathcal{O}v, v >= \varepsilon < v, v >$, where $< v, v >$ (> 0) is real and also $< \mathcal{O}v, v >$ is real by proposition 7.2.12. We conclude from this ε is real.

Proposition(8.3.2). *Suppose that v is an eigenvector of a hermitian operator $\mathcal{O}$ of V; then for any vector w orthogonal to v, v is also orthogonal to $\mathcal{O}w$. Also if a subspace S satisfies $\mathcal{O}(S) \subset S$, then $\mathcal{O}(S^{\perp}) \subset S^{\perp}$ holds, i.e., $S^{\perp}$ also becomes $\mathcal{O}$-invariant.*

Proof. Very similar to that of proposition 8.2.1.

Proposition(8.3.3). *(Spectral basis theorem)*

For any hermitian operator $\mathcal{O}$, we have an orthogonal basis of V consisting of eigenvectors of $\mathcal{O}$ and so an orthonormal basis of V exists, where V is a finite dimensional vector space over $\mathbb{C}$ with a positive definite hermitian product.

Proof. Just the same as that of proposition 8.2.3.

Finally we should note that the following assertion is very similar to that of proposition 8.2.4, but its proof is slightly different from the corresponding one in view of hermitian form.

Proposition(8.3.4). *Suppose that M is an $\ell \times \ell$ complex hermitian matrix; we then have a complex unitary matrix u satisfying $u^{-1}Mu = {}_*uMu =$ a diagonal matrix.*

Proof. Let us view M as an associated matrix with an operator $\mathcal{O} : {}_t\mathbb{C}^\ell \longrightarrow {}_t\mathbb{C}^\ell$. By the spectral basis theorem we have an orthonormal basis (of ${}_t\mathbb{C}^\ell$)$\widetilde{\mathcal{B}}$ consisting of eigenvectors of ${}_t\mathbb{C}^\ell$, say $\widetilde{\mathcal{B}} = \{e_1, \cdots, e_\ell\}$.

Let $\mathcal{B}$ be the standard elementary basis of ${}_t\mathbb{C}^\ell$. We then have $A^{\mathcal{B}}_{\widetilde{\mathcal{B}}}(id) \circ A^{\mathcal{B}}_{\mathcal{B}}(\mathcal{O}) \circ A^{\widetilde{\mathcal{B}}}_{\mathcal{B}}(id) = A^{\widetilde{\mathcal{B}}}_{\widetilde{\mathcal{B}}}(\mathcal{O})$, i.e., $A^{\mathcal{B}}_{\widetilde{\mathcal{B}}}(id) \cdot M \cdot A^{\widetilde{\mathcal{B}}}_{\mathcal{B}}(id) =$ a diagonal matrix over $\mathbb{C}$. We put $u = A^{\widetilde{\mathcal{B}}}_{\mathcal{B}}(id)$ and claim that it is complex unitary. In fact $id(e_i) = e_i = \sum_{j=1}^{\ell} c_{ji}u_j$ for $i = 1, \cdots, \ell$ and for some $c_{ji} \in \mathbb{C}$. Since $< e_i, e_k >= \sum_{j=1}^{\ell} c_{ji}\bar{c_{jk}} = 0$ if $i \neq k$ and $< e_i, e_i >= \sum_{j=1}^{\ell} c_{ji}\bar{c_{ji}} = \sum_{j=1}^{\ell} |c_{ji}|^2 = 1$, we have $\overline{{}_tA^{\widetilde{\mathcal{B}}}_{\mathcal{B}}(id)} \cdot A^{\widetilde{\mathcal{B}}}_{\mathcal{B}}(id) = {}_t(\bar{c_{ji}})(c_{ji}) = I_\ell$. Hence

$$ {}_*A^{\widetilde{\mathcal{B}}}_{\mathcal{B}}(id) = \overline{{}_tA^{\widetilde{\mathcal{B}}}_{\mathcal{B}}(id)} = A^{\widetilde{\mathcal{B}}}_{\mathcal{B}}(id)^{-1} $$

holds showing that $A^{\widetilde{\mathcal{B}}}_{\mathcal{B}}(id)$ is complex unitary. So $u := A^{\widetilde{\mathcal{B}}}_{\mathcal{B}}(id)$ finishes the proof.

Exercises of § 8-3

[1] If u is complex unitary, show that so is u^{-1}.

[2] For any $\ell \times \ell$ complex hermitian matrix M, show that an orthonormal basis of $\mathbb{C}^\ell$ consisting of eigenvectors of M exists.

[3] Let V be designated as in this section. We say that an operator $\mathcal{O} : V \longrightarrow V$ is *normal* if $\mathcal{O}_*\mathcal{O} = {}_*\mathcal{O}\mathcal{O}$.

(i) State and prove a spectral basis theorem for a normal operator $\mathcal{O}$.

(ii) If two normal operators $\mathcal{O}, \mathcal{P}$ satisfy $\mathcal{O}\mathcal{P} = \mathcal{P}\mathcal{O}$, prove that $\mathcal{O}\mathcal{P}$ is also normal.

[4] Show that ${}_*\mathcal{O}$ of a normal operator $\mathcal{O}$ is also a normal operator.

§ 8-4. Complex unitary operator

In this section we assume that V is a finite dimensional vector space over $\mathbb{C}$ with a positive definite hermitian form.

Proposition(8.4.1). *Suppose that $\mathcal{O}$ is a (complex) unitary operator of V as above. If S is an $\mathcal{O}$-invariant subspace of V, then so is $S^{\perp}$.*

Proof. Note for a moment that $\mathcal{O}$ is invertible because ${}_*\mathcal{O} = \mathcal{O}^{-1}$. Now $\forall v \in S$, $\forall w \in S^{\perp}$, $< v, \mathcal{O}w >=< {}_*\mathcal{O}v, w >=< \mathcal{O}^{-1}v, w >= 0$. Hence $S^{\perp}$ is also $\mathcal{O}$-invariant.

It is not amazing that there may be also a spectral basis for a complex unitary operator.

Proposition(8.4.2). *(***Spectral basis theorem***)*

Suppose that $\mathcal{O}$ is a unitary operator of a finite dimensional $\mathbb{C}$-vector space V with a positive definite hermitian product; we then have an orthogonal and so an orthonormal basis consisting of eigenvectors of $\mathcal{O}$.

Proof. Evidently we have an eigenvalue ε of $\mathcal{O}$ from proposition 8.1.9, because V is a $\mathbb{C}$-vector space. So we let $\mathcal{O}v_1 = \varepsilon v_1$ for some eigenvector v_1 of $\mathcal{O}$. Put $S := \mathbb{C}v_1$. We thus have that $S^{\perp}$ is also $\mathcal{O}$-invariant according to proposition 8.4.1.

Now applying the induction hypothesis to $S^{\perp}$ yields a spectral basis $\{v_2, \cdots, v_\ell\}$ of $\mathcal{O}$-invariant subspace $S^{\perp}$. After all $\{v_1, v_2, \cdots, v_\ell\}$ should be a spectral basis of V for $\mathcal{O}$.

Remark(8.4.3). Also in this situation, we may have the similar assertion to proposition 8.3.4.

Exercises of § 8-4

[1] Show that any eigenvalue of a complex unitary operator is nonzero, where the vector space is given as in this section.

[2] Show that $\begin{pmatrix} i & i \\ i & -i \end{pmatrix}$ is not a unitary matrix.

[3] Show that $diag(e^{i\theta_1}, \cdots, e^{i\theta_n}) \forall \theta_i \in \mathbb{R}$ is a complex unitary matrix.

[4] Find the spectral basis of the unitary matrix $diag(e^{i\theta_1}, \cdots, e^{i\theta_n})\ \forall \theta_i \in \mathbb{R}$.

§ 8-5. Real unitary operator

Unfortunately we cannot say in general that a real unitary operator has a spectral basis. But we shall do our best in this direction. Let V be a finite dimensional vector space with a positive definite scalar product. We consider the decomposition of V in what follows.

Proposition(8.5.1). *Let V be as above and let $\mathcal{O}$ be a real unitary operator of V; then we may decompose V into a direct sum of $\mathcal{O}$-invariant subspaces V_i such as $V = V_1 \oplus \cdots \oplus V_k$, where each V_i has 1 or 2 as its dimension and $< V_i, V_j >= 0$ if $i \neq j$ for $1 \leq i, j \leq k$.*

Proof. Because V has an orthonormal basis, we may regard V as $\mathbb{R}^\ell$ with respect to this basis. Let A be the associated matrix with $\mathcal{O}$ with respect to this basis. Considering $A : {}_t\mathbb{C}^\ell \longrightarrow {}_t\mathbb{C}^\ell$ as a natural operator, we have ${}_t\bar{A} = A^{-1}$ since $\mathcal{O}$ is real unitary.

Because there is a nonzero eigenvector of A in ${}_t\mathbb{C}^\ell$ with some eigenvalue ε, $AZ = \varepsilon Z$ happens, so that $\|AZ\| = \|Z\| = |\varepsilon| \cdot \|Z\|$ yields $\varepsilon = e^{i\theta}$ for some $\theta \in \mathbb{R}$. If we put $Z = X + iY$ with some $X, Y \in {}_t\mathbb{R}^\ell$, we get 2 kinds of A-invariant subspaces relative to ε as follows :

(i) $\varepsilon = e^{i\theta}$ is real; then $e^{i\theta} = \pm 1$ gives rise to $AX = \varepsilon X$ and $AY = \varepsilon Y$. Because X and Y are not all zero, an eigenvector v_1 for $\mathcal{O} : V \longrightarrow V$ has been found. Putting $V_1 = \mathbb{R}v_1$, we get $V_1 \oplus V_1^\perp = V$, where $V_1^\perp$ is also $\mathcal{O}$-invariant by virtue of the similar reason as in proposition 8.4.1. Now applying the induction hypothesis to $V_1^{\perp}$ gives the result as required.

(ii) $\varepsilon = e^{i\theta} = cos\theta + isin\theta \neq cos\theta$; then we have another different eigenvalue $\bar{\varepsilon}$ since $A\bar{Z} = \bar{\varepsilon}\bar{Z}$. We also have

$$\begin{aligned} AZ &= AX + iAY = \varepsilon(X + iY) \\ &= \{(cos\theta)X - (sin\theta)Y\} + i\{(sin\theta)X + (cos\theta)Y\} \end{aligned}$$

, so that it gives

$$\begin{cases} AX = (cos\theta)X - (sin\theta)Y \\ AY = (sin\theta)X + (cos\theta)Y. \end{cases}$$

We claim that X and Y are linearly independent. Indeed if $rX = Y$ for some $r \in \mathbb{R}$, then from the above equations we obtain

$$\begin{cases} rAX = (cos\theta)rX - (sin\theta)r^2X \\ ArX = (sin\theta)X + (cos\theta)rX. \end{cases}$$

Subtracting one from the other equation gives $(sin\theta)(1+r^2)X = 0$. So if $X = 0$, then $Y = 0$ gives $Z = 0$, a contradiction. If $sin\theta = 0$, then ε and $\bar{\varepsilon}$ are not different eigenvalues, also a contradiction.

Next let S be the subspace of V generated by vectors X, Y (over $\mathbb{R}$) in V; then by the above argument we see that S is a 2-dimensional $\mathcal{O}$-invariant subspace of V. Hence we get the decomposition $V = S \oplus S^{\perp}$, where $S^{\perp}$ is also an $\mathcal{O}$-invariant subspace. Now applying the induction hypothesis to $S^{\perp}$ gives the final decomposition as required.

Corollary(8.5.2). *Let $\mathcal{O}$ and V be as before. We then have a basis $\mathcal{B}$ of V so that $A^{\mathcal{B}}_{\mathcal{B}}(\mathcal{O})$ is a matrix of the form*

$$\begin{pmatrix} A_1 & 0 & \cdots & 0 \\ 0 & A_2 & \cdots & 0 \\ \vdots & \vdots & \ddots & \vdots \\ 0 & 0 & \cdots & A_k \end{pmatrix}$$

with each A_i a 1×1 matrix or a 2×2 matrix of the following prototypes:

$$(1), \quad (-1), \quad \begin{pmatrix} cos\theta & sin\theta \\ -sin\theta & cos\theta \end{pmatrix}.$$

Hence $\mathcal{O}$ acts on each component of the above decomposition of the preceding proposition as either an identity id or a reflection- id or a rotation by turns.

Proof. We just conclude the corollary from the above proposition.

Exercises of § 8-5

[1] If the vector space is given as in this section, verify that any eigenvalue of a real unitary operator is nonzero.

[2] Prove or disprove that $\begin{pmatrix} cos\theta & sin\theta \\ -cos\theta & sin\theta \end{pmatrix}$ is real unitary.

[3] Find the eigenvalues of $\begin{pmatrix} cos\theta & -sin\theta \\ sin\theta & cos\theta \end{pmatrix}$ noting that it is real unitary.

CHAPTER 9. POLYNOMIALS AND OPERATORS

For a general decomposition of a vector space into $\mathcal{O}$-invariant subspaces, we need some knowledges from polynomial theory. In particular Hamilton-Cayley's theorem or Euclidean algorithm is critical in some problems.

§ 9-1. Polynomials of matrices and operators

Let $F[\mathbb{X}]$ denote the collection of all polynomials with coefficients in F. We already know the well-defined operations : $\forall p(\mathbb{X}), q[\mathbb{X}] \in F[\mathbb{X}], (p+q)(\mathbb{X}) := p(\mathbb{X}) + q(\mathbb{X})$ and $(pq)(\mathbb{X}) := p(\mathbb{X}) \cdot q(\mathbb{X})$.

Definition(9.1.1). Let M be any $\ell \times \ell$ matrix over F and let $p(\mathbb{X}) = f_0 + f_1\mathbb{X} + \cdots + f_m\mathbb{X}^m$; we then define $p(M) := f_0 I_\ell + f_1 M + \cdots + f_m M^m \in M_{\ell\times\ell}(F)$. For any operator $\mathcal{O}$ of a vector space V over F, we also define $p(\mathcal{O}) := f_0 I + f_1\mathcal{O} + \cdots + f_m\mathcal{O}^m$ which is again an operator.

Proposition(9.1.2). *With notations as above we have*

$$(p+q)(M) = p(M) + q(M)$$
$$(pq)(M) = p(M)q(M)$$

which are also well-defined. Also just the same for $\mathcal{O}$ instead of M.

Proof. For any operators $\mathcal{O}_1, \mathcal{O}_2, \mathcal{O}_3$ (matrices M_1, M_2, M_3 respectively) we have the distributive law $(\mathcal{O}_1 + \mathcal{O}_2) \circ \mathcal{O}_3 = \mathcal{O}_1 \circ \mathcal{O}_3 + \mathcal{O}_2 \circ \mathcal{O}_3$ and also the commutative laws $\begin{pmatrix} \mathcal{O}_1^i \circ \mathcal{O}_1^j = \mathcal{O}_1^j \circ \mathcal{O}_1^i \\ \mathcal{O}_1 + \mathcal{O}_2 = \mathcal{O}_2 + \mathcal{O}_1 \end{pmatrix}$. So our assertion is clear. Also we have the similar assertion for M.

Proposition(9.1.3). *For any $M \in M_{\ell \times \ell}(F)$, there exists a nonzero polynomial $p(\mathbb{X}) \in F[\mathbb{X}]$ satisfying $p(M) = 0$. Also it holds for any operator $\mathcal{O} : V \longrightarrow V$ if V is finite dimensional over F.*

Proof. Since $M_{\ell \times \ell}(F)$ is a vector space over F with its dimension ℓ^2, we see that $I, M, M^2, \cdots, M^m$ with $m \geq \ell^2$ are linearly dependent. So for some coefficients $f_i \in F$ (not all zero) we have $f_0 I + f_1 M + \cdots + f_m M^m = 0$. Now letting $p(\mathbb{X}) := f_0 + f_1 \mathbb{X} + \cdots + f_m \mathbb{X}^m$ completes the proof.

Exercises of § 9-1

[1] Let $p(M) = a_n M^n + \cdots + a_0 I$, $q(M) = b_m M^m + \cdots + b_0 I$ for $\forall M \in M_{\ell \times \ell}(F)$. Show that $(pq)(M) = c_0 I + c_1 M + \cdots + c_{m+n} M^{m+n}$, where $c_j = \sum_{i=0}^{j} a_i b_{j-i}$.

[2] Solve the analogue of the above [1] with M replaced by an operator $\mathcal{O}$ of a finite dimensional vector space V over F.

[3] If $\mathcal{P}$ is an invertible operator and $\mathcal{O}$ is an operator of a vector space over F, show that $p(\mathcal{P}^{-1}\mathcal{O}\mathcal{P}) = \mathcal{P}^{-1}p(\mathcal{O})\mathcal{P}$ for any polynomial $p(\mathbb{X})$.

§ 9-2. Hamilton-Cayley's theorem

It is not obvious that we may actually find out a polynomial $p(\mathbb{X})$ such that $p(M) = 0$ for a given square matrix M even though we admit proposition 9.1.3. We introduce the following Hamilton-Cayley's theorem.

Theorem(9.2.1). *For any square matrix* $M \in M_{\ell\times\ell}(F)$, $p_M(\mathbb{X})$ *satisfies* $p_M(M) = 0$.

Proof. Let us call a matrix with entries in $F[\mathbb{X}]$ an $\mathbb{X}$-matrix, while we call a matrix $\in M_{\ell\times\ell}(F)$ a matrix of scalars. If we consider the multiples $\mathbb{X}M, \mathbb{X}^2M, \cdots, \mathbb{X}^mM$ in the natural manner, these are also $\mathbb{X}$-matrices.

In any nonzero $\mathbb{X}$-matrix A we let h denote the highest degree of any nonzero polynomial which is an entry of A. We may thus write A in terms of matrices $A^{(i)}$ of scalars as an obvious sum $A = A^{(0)} + \mathbb{X}A^{(1)} + \cdots + \mathbb{X}^hA^{(h)}$.

Remembering the familar factorization $f^i - \mathbb{X}^i = (f^{i-1} + f^{i-2}\mathbb{X} + \cdots + \mathbb{X}^{i-1})(f - \mathbb{X})$ for $\forall f \in F$ and for $i = 1, 2, 3, \cdots$, we get an analog of this such as $M^i - \mathbb{X}^iI = (M^{i-1} + \mathbb{X}M^{i-2} + \cdots + \mathbb{X}^{i-1}I)(M - \mathbb{X}I)$. If we let $L^{(i)}$ designate the $\mathbb{X}$-matrix $M^{i-1} + \mathbb{X}M^{i-2} + \cdots + \mathbb{X}^{i-1}I$ with $L^{(0)} = 0$ and $L^{(1)} = I$, we have $M^i - \mathbb{X}^iI = L^{(i)}(M - \mathbb{X}I)$ obviously for $i = 0, 1, 2, \cdots$.

Next we consider the characteristic polynomial $D(M - \mathbb{X}I) = p_M(\mathbb{X}) = f_0 + f_1\mathbb{X} + \cdots + f_\ell\mathbb{X}^\ell$ of M with $f_\ell = \pm 1$. Because $adj(M - \mathbb{X}I_\ell) \cdot (M - \mathbb{X}I_\ell) = p_M(\mathbb{X})I_\ell$ and $M^i - \mathbb{X}^iI_\ell = L^{(i)}(M -$

$\mathbb{X}I_\ell)$, we obtain

$$p_M(M) = \sum_{i=0}^{\ell} f_i M^i = \sum_{i=0}^{\ell} f_i \mathbb{X}^i I_\ell + \sum_{i=0}^{\ell} f_i L^{(i)}(M - \mathbb{X}I_\ell).$$

Hence from the above equations, we have

$$\begin{aligned} p_M(M) &= p_M(\mathbb{X})I_\ell + \sum_{i=0}^{\ell} f_i L^{(i)}(M - \mathbb{X}I_\ell) \\ &= adj(M - \mathbb{X}I_\ell) \cdot (M - \mathbb{X}I_\ell) + \sum_{i=0}^{\ell} f_i L^{(i)}(M - \mathbb{X}I_\ell) \\ &= \{adj(M - \mathbb{X}I_\ell) + \sum_{i=0}^{\ell} f_i L^{(i)}\}(M - \mathbb{X}I_\ell). \end{aligned}$$

If $adj(M - \mathbb{X}I_\ell) + \sum_{i=0}^{\ell} f_i L^{(i)}$ is nonzero, it then gives rise to a contradiction since the left-hand side of the last equation is a matrix of scalars while the right-hand side is an $\mathbb{X}$-matrix which is not a matrix of scalars. Hence $p_M(M) = 0$.

Corollary(9.2.2). *The polynomial of least degree in $F[\mathbb{X}]$ satisfied by an $\ell \times \ell$ square matrix M like a root divides the characteristic polynomial. So its degree is at most ℓ.*

Proof. By the Division algorithm appearing in the next section and by the above proposition, the assertion is evident. Indeed let J be the set of all polynomials $p(\mathbb{X}) \in F[\mathbb{X}]$ satisfying $p(M) = 0$; evidently $\forall p, q \in J, p + q \in J$, and $\forall r \in F[\mathbb{X}], \forall p \in J, rp \in J$.

In this case we say that J is an *ideal* of $F[\mathbb{X}]$. Let $g \in J$ be a nonzero element of least degree in J. By the Division algorithm,

$\forall p(\in J) = qg + r$ for some $q \in F[\mathbb{X}]$ with $deg\ r < deg\ g$, where we define $deg\ 0 = -\infty$. But then $r = p - qg = p + (-q)g \in J$, so that we have $r = 0$. Now from the preceding proposition we conclude our assertion.

Definition(9.2.3). The polynomial $q(\mathbb{X})$ of least degree satisfied by an $\ell \times \ell$ square matrix M, i.e., $g(M) = 0$, as above is called the *minimal polynomial* of M. The above J is called an *ideal* of $F[\mathbb{X}]$. Explicitly if $\forall p, q \in J \subset F[\mathbb{X}], p + q \in J$ and if $\forall r \in F[\mathbb{X}], \forall p \in J, rp \in J$, then we call J an ideal of $F[\mathbb{X}]$.

Remark(9.2.4). We also call the above $g(\mathbb{X})$ in the proof a generator of the ideal J. We shall study this more in § 9-4.

Exercises of § 9-2

[1] Let V be a finite dimensional vector space over F and let V have a basis $\mathcal{B}$ consisting of eigenvectors of an operator $\mathcal{O} : V \longrightarrow V$. If $\varepsilon_1, \cdots, \varepsilon_\ell$ are their corresponding eigenvalues, show that $p_M(\mathbb{X}) = (\mathbb{X} - \varepsilon_1) \cdots (\mathbb{X} - \varepsilon_\ell)$, where $M = A_{\mathcal{B}}^{\mathcal{B}}(\mathcal{O})$.

[2] In the notation of [1], verify that $p_M(\mathcal{O}) = 0$.

[3] Show that if $p_M(\mathbb{X}) = \mathbb{X}^2 - 1$ for a 2×2 square matrix M, then M is similar to a diagonal matrix, where the characteristic of $F \neq 2$.

[4] What is the minimal polynomial of a diagonal matrix ?

§ 9-3. Division algorithm and Euclidean algorithm

In order to calculate the greatest common divisor (abb.g.c.d.) or the least common multiple (abb.l.c.m.) of two polynomials in $F[\mathbb{X}]$, we should make use of the well-known division algorithm.

Proposition(9.3.1). *If $p(\mathbb{X}), q(\mathbb{X}) \in F[\mathbb{X}]$ with deg $q(\mathbb{X}) \geq 0$; we then have $\widetilde{q}(\mathbb{X})$ and $r(\mathbb{X})$ uniquely such that $p(\mathbb{X}) = \widetilde{q}(\mathbb{X})q(\mathbb{X})$ $+r(\mathbb{X})$ with deg $r(\mathbb{X}) <$ deg $q(\mathbb{X})$.*

Proof. If $deg\, p(\mathbb{X}) < deg\, q(\mathbb{X})$, then putting $\widetilde{q}(\mathbb{X}) = 0$ and $r(\mathbb{X}) = p(\mathbb{X})$ finishes the proof. Otherwise let $p(\mathbb{X}) = a_0 + a_1\mathbb{X} + \cdots + a_m\mathbb{X}^m$ and $q(\mathbb{X}) = b_0 + b_1\mathbb{X} + \cdots + b_n\mathbb{X}^n$ with $a_i, b_j \in F$ and $a_m, b_n \neq 0$. Because $p(\mathbb{X}) - \frac{a_m}{b_n}\mathbb{X}^{m-n}q(\mathbb{X})$ has its degree$\leq m-1$, applying induction on m finishes the proof. Namely

$$p(\mathbb{X}) - \frac{a_m}{b_n}\mathbb{X}^{m-n}q(\mathbb{X}) = \widetilde{r}(\mathbb{X})q(\mathbb{X}) + r(\mathbb{X})$$

for some $\widetilde{r}(\mathbb{X}), r(\mathbb{X}) \in F[\mathbb{X}]$ with $deg\, q(\mathbb{X}) > deg\, r(\mathbb{X})$. So $p(\mathbb{X}) = \{\widetilde{r}(\mathbb{X}) + \frac{a_m}{b_n}\mathbb{X}^{m-n}\}q(\mathbb{X}) + r(\mathbb{X})$ is attained as required. Uniqueness is evident in view of degree.

Remark(9.3.2). So as to find the g.c.d of $p_0(\mathbb{X}), p_1(\mathbb{X}) \in F[\mathbb{X}]$, first divide p_0 by p_1, as in the division algorithm, to get a remainder p_2 of smaller degree, then divide p_1 by the remainder p_2 getting a new remainder p_3, and so forth until some remainder, say p_{n+1} finally becomes zero which sure exists because the degrees of the remainders decrease.

We may express this process as a table of equations with successive quotient polynomials q_i:

$$
\begin{aligned}
p_0 &= q_1 p_1 + p_2 \quad \text{with} \quad deg\ p_1 > deg\ p_2, \\
p_1 &= q_2 p_2 + p_3 \quad \text{with} \quad deg\ p_2 > deg\ p_3, \\
&\vdots \qquad\qquad\qquad\qquad\qquad \vdots \\
p_{n-2} &= q_{n-1} p_{n-1} + p_n \quad \text{with} \quad deg\ p_{n-1} > deg\ p_n, \\
p_{n-1} &= q_n p_n,
\end{aligned}
$$

which is the so-called Euclidean Algorithm for two polynomials $p_0(\mathbb{X}), p_1(\mathbb{X}) \in F[\mathbb{X}]$. After we have learned the unique factorization property in § 9-5, we shall see easily that p_n is just g.c.d. of p_0 and p_1.

Corollary(9.3.3). *If* $\widetilde{r} \in F$ *a root of a nonzero polynomial* $p(\mathbb{X}) \in F[\mathbb{X}]$, *then* $p(\mathbb{X}) = (\mathbb{X} - \widetilde{r})q(\mathbb{X})$ *for some* $q(\mathbb{X}) \in F[\mathbb{X}]$.

Proof. If we divide p by $\mathbb{X} - \widetilde{r}$, we get $p(\mathbb{X}) = q(\mathbb{X})(\mathbb{X} - \widetilde{r}) + r(\mathbb{X})$ with $r(\mathbb{X}) \in F$ by virtue of the division algorithm. Now $p(\widetilde{r}) = r(\mathbb{X}) = 0$ by the hypothesis.

Note that we have used before the following property which we believe $\mathbb{C}$ has . See appendix for the proof.

Corollary(9.3.4). *Suppose that a field* F *has the property that every nonconstant polynomial* $p(\mathbb{X}) \in F[\mathbb{X}]$ *has a root in* F; *we then have* $p(\mathbb{X}) = a_m(\mathbb{X} - r_1) \cdots (\mathbb{X} - r_m)$ *for some* $a_m, r_1, \cdots, r_m \in F$ *which may not be distinct.*

Proof. Straightforward from corollary 9.3.3.

Definition(9.3.5). A field F having the property as in corollary 9.3.4 is called *algebraically closed* in the literature.

Exercises of § 9-3

[1] Confirm the Euclidean algorithm in $\mathbb{Z}$ and express the indicated g.c.d. as a linear combination of the two given integers :

(a) g.c.d. (254,810) (b) (4867,8414) (c) (15,33).

[2] Do the same jobs in $\mathbb{Q}(\mathbb{X})$ for the following as in [1]:

(a) g.c.d.$(x^3+1, x+1)$ (b) g.c.d.$(x^{33}-1, x^{18}-1)$

(c) g.c.d.$(x^4+x^3+2x^2+x+1, x^3-1)$.

[3] Suppose that $p(\mathbb{X}), q(\mathbb{X})$ have leading coefficients one and integer coefficients. If we use the division algorithm for $p(\mathbb{X}), q(\mathbb{X})$, show that all related polynomials to these algorithms have integer coefficients too.

[4] Offer the proof of corollary 9.3.4 in detail.

[5] Use corollary 9.3.4 to factorize $p(\mathbb{X}) = x^3-6x^2+11x-6 \in \mathbb{Q}[\mathbb{X}]$.

§ 9-4. GCD and LCM

As usual, we say $p(\mathbb{X})$ divides $q(\mathbb{X})$ denoted by $p|q$ whenever $q(\mathbb{X}) = \widetilde{q}(\mathbb{X}) \cdot p(\mathbb{X})$ for some $\widetilde{q}(\mathbb{X}) \in F[\mathbb{X}]$. Also we say $g(\mathbb{X})$ is the g.c.d. of $p(\mathbb{X}), q(\mathbb{X})$ if $g|p$ and $g|q$, and if in addition $h|p$ and $h|q$ implies $h|g$. We also understand l.c.m. as usual.

Definition(9.4.1). Given polynomials $g_1, \cdots, g_\ell \in F[\mathbb{X}]$, we let J be the collection of all polynomials of the form $f_1 g_1 + \cdots + f_\ell g_\ell$ with $f_i \in F[\mathbb{X}]$. We easily see that J is an ideal of $F[\mathbb{X}]$. We call such $g_1, \cdots, g_\ell$ *generators* of J and say that $g_1, \cdots, g_\ell$ generate the ideal J.

Proposition(9.4.2). *Any ideal of $F[\mathbb{X}]$ has a single generator uniquely determined up to scalars.*

Proof. Just the recapitulation of the relevant part in the proof of corollary 9.2.2.

Proposition(9.4.3). *For nonzero polynomials $g_1, \cdots, g_\ell \in F[\mathbb{X}]$, we let g be a generator for the ideal generated by $g_1, \cdots, g_\ell$. We then have $g = g.c.d.(g_1, \cdots, g_\ell)$.*

Proof. Let J be the ideal generated by $g_i, i = 1, \cdots, \ell$. Since $g_i = 0g_1 + \cdots + g_i + 0g_{i+1} + \cdots + 0g_\ell = \sum_{j=1}^{\ell} \delta_{ij} g_j$ holds, we have $g|g_i, i = 1, \cdots, \ell$.

Next if $h|g_i$, then $g_i = f_i h$ for some $f_i \in F[\mathbb{X}], i = 1, \cdots, \ell$. Hence h divides any element in J; in particular $h|g$ completes the proof.

Definition(9.4.4). We understand that a generator is uniquely determined if we choose it with leading coefficient 1. In this case we say that a generator is *uniquely determined up to constants.* We also say that $g_1, \cdots, g_\ell$ are *relatively prime* if their g.c.d. is just 1.

Exercises of § 9-4

[1] List all divisors of $\mathbb{X}^7 - \mathbb{X} \in \mathbb{Q}[\mathbb{X}]$ up to constants.

[2] Find a generator of the ideal generated by $\mathbb{X}+1,\ \mathbb{X}^2-1$.

[3] Verify $(\mathbb{X}+1)|(\mathbb{X}^m+1)$ if m is odd.

[4] Show $(\mathbb{X}-1)|(\mathbb{X}^m-1)$ for $\forall m \in \mathbb{N}$.

[5] What is the ideal generated by relatively prime polynomials ?

§ 9-5. Polynomials' unique factorization

We should discern a product of polynomials from a single polynomial which can't be factored.

Definition(9.5.1). If a nonzero polynomial $p(\mathbb{X})$ has only 1 and p itself as its divisors up to constants, we then call p *irreducible* in $F[\mathbb{X}]$.

Example(9.5.2). Let $p(\mathbb{X}) = \mathbb{X}^2 + 1$. Then $p(\mathbb{X}) \in \mathbb{R}[\mathbb{X}]$ is irreducible, while $p(\mathbb{X}) \in \mathbb{C}[\mathbb{X}]$ is reducible in the sense that $\mathbb{X}^2 + 1 = (\mathbb{X} + i)(\mathbb{X} - i)$ over $\mathbb{C}$.

Proposition(9.5.3). *Suppose that p is an irreducible polynomial $\in F[\mathbb{X}]$ and that $p|qr$ for some nonzero polynomial $q, r \in F[\mathbb{X}]$. We then have $p|q$ or $p|r$ happens.*

Proof. If p does not divide q, then irreducible p and q are relatively prime. Hence from proposition 9.4.3, we obtain $1 = \widetilde{p} \cdot p + \widetilde{q} \cdot q$ for some $\widetilde{p}, \widetilde{q} \in F[\mathbb{X}]$. Multiplying both sides by r gives $r = r\widetilde{p} \cdot p + \widetilde{q} \cdot qr$, so $p|r$ since $p|qr$.

Proposition(9.5.4). *Any nonzero polynomial of $F[\mathbb{X}]$ may be uniquely expressed as a product of irreducible polynomials up to order and nonzero constants.*

Proof. We shall prove the existence at first. Let $p(\mathbb{X})$ be any nonzero polynomial of $F[\mathbb{X}]$. We may prove by induction on *deg* $p(\mathbb{X})$ that $\forall p(\mathbb{X})$ has a product of irreducible polynomials. If

$deg\ p = 0$ or 1, then the existence of such a product is trivial. Next let $deg\ p > 1$. If it is irreducible, then we are done.

Otherwise p has two proper factors giving $p = q \cdot r$ for some $q, r \in F[\mathbb{X}]$ of degree≥ 1. Since $deg\ q, deg\ r < deg\ p$, we see that q, r are products of irreducible polynomials in $F[\mathbb{X}]$. So if we adapt the induction hypothesis to q and r, we can assure the existence of a product of irreducible polynomials with respect to $\forall p(\mathbb{X}) \in F[\mathbb{X}]$.

Next for the uniqueness proof, we assume that $\exists$ two products of irreducible polynomials like $p(\mathbb{X}) = i_1(\mathbb{X}) \cdots i_m(\mathbb{X}) = j_1(\mathbb{X}) \cdots j_\ell(\mathbb{X})$. After some permutation of indices we may also assume that $i_1 = f_1 j_1$ for some $f_1 \in F$ because of irreducibility and proposition 9.5.3. Now the inductive repetition of such an argument renders $i_k = f_k j_k\ \forall k = 1, \cdots, m$ after cancelling j_1, completing the uniqueness proof up to scalars.

Corollary(9.5.5). *Any $p(\mathbb{X})$ of degree≥ 1 in $\mathbb{C}[\mathbb{X}]$ has a unique factorization up to permutation like $p(\mathbb{X}) = c(\mathbb{X} - r_1) \cdots (\mathbb{X} - r_\ell)$ for some $c \in \mathbb{C}$ and some roots $r_i \in \mathbb{C}$.*

Proof. Immediate conclusion from the proposition.

Definition(9.5.6). If $p(\mathbb{X}) = (\mathbb{X}-r)^m q(\mathbb{X})$ in $F[\mathbb{X}]$ and $(\mathbb{X}-r) \nmid q(\mathbb{X})$, then we call m the *multiplicity* of $(\mathbb{X} - r)$(or simply r) in $p(\mathbb{X})$.

Exercises of § 9-5

[1] If *deg* $p(\mathbb{X}) = 3$ in $F[\mathbb{X}]$ and $p(\mathbb{X})$ is not irreducible, show then that some $r \in F$ satisfies $p(\mathbb{X}) = 0$.

[2] For any $p(\mathbb{X}) \in F[\mathbb{X}]$, we define $\frac{d}{d\mathbb{X}}p(\mathbb{X}) = p'(\mathbb{X})$ as follows. If $p(\mathbb{X}) = f_0 + f_1\mathbb{X} + \cdots + f_\ell\mathbb{X}^\ell \in F[\mathbb{X}]$, then $\frac{d}{dx}p(\mathbb{X}) = p'(\mathbb{X}) = f_1 + 2f_2\mathbb{X} + \cdots + (\ell - 1)f_{\ell-1}\mathbb{X}^{\ell-2} + \ell f_\ell\mathbb{X}^{\ell-1} \in F[\mathbb{X}]$. In this case verify the indicated problems :

(i) $\forall p(\mathbb{X}), q(\mathbb{X}) \in F[\mathbb{X}], (p+q)' = p' + q'$, and $(pq)' = p'q + pq'$,

(ii) Let $p(\mathbb{X}) \in F[\mathbb{X}]$ be of degree≥ 1 and have a root $r \in F$. The multiplicity of r in $p(\mathbb{X})$ is greater than 1 if and only if $p'(r) = 0$. (Hint : consider integration.)

CHAPTER 10. JORDAN NORMAL FORM

We should do our best in order to make any operator or matrix as simple as possible in the sense that diagonal ones are the simplest. So we have to find out or assure some basis of a vector space with respect to which the given operator is simplest. In this sense we might as well say that Jordan normal form is the final goal of all previous chapters. As is often the case, the last goal looks hard seemingly.

§ 10-1. Decomposition of vector spaces

Let S be a subset of an F-vector space V and $\widetilde{S}$ be a set of certain operators of V. As usual we shall say that S is $\widetilde{S}$-invariant if $\forall \mathcal{O} \in \widetilde{S}, \forall v \in S, \mathcal{O}v \in S$.

Example(10.1.1). Eigenspaces V_ε of an operator $\mathcal{O}$ of V over F are evidently $\mathcal{O}$-invariant. Of course V itself and (0) are $\mathcal{O}$-invariant trivially.

Proposition(10.1.2). *If $p(\mathbb{X}) \in F[\mathbb{X}]$ and $\mathcal{O}$ is an operator of an F-vector space V, then $Ker\ p(\mathcal{O})$ becomes $\mathcal{O}$-invariant.*

Proof. We have only to show that $\forall v \in Ker\ p(\mathcal{O})$, $\mathcal{O}v \in Ker\ p(\mathcal{O})$. It is, however, easily proved if we consider $p(\mathcal{O})\mathcal{O} = \mathcal{O}p(\mathcal{O})$. Indeed $p(\mathcal{O})\mathcal{O}v = \mathcal{O}p(\mathcal{O})v = \mathcal{O}(0) = 0$.

Proposition(10.1.3). *Suppose that $p(\mathbb{X}) \in F[\mathbb{X}]$ is a nonconstant polynomial and $p(\mathbb{X}) = p_1(\mathbb{X})p_2(\mathbb{X})$ for some relatively prime polynomials p_1, p_2. Suppose further that $p(\mathcal{O}) = 0$ for an operator $\mathcal{O}$ of V over F.*

We then have $V = Ker\ p_1(\mathcal{O}) \oplus Ker\ p_2(\mathcal{O})$.

Proof. We know from proposition 9.4.3 that $1 = q_1p_1 + q_2p_2$ for some polynomials q_1, q_2 in $F[\mathbb{X}]$, whence $\forall v \in V, v = q_1(\mathcal{O})p_1(\mathcal{O})$ $v + q_2(\mathcal{O})p_2(\mathcal{O})v$ holds. Since $p_1(\mathcal{O})v \in Ker\ p_2(\mathcal{O})$ and $p_2(\mathcal{O})v \in Ker\ p_1(\mathcal{O})$, we easily see that $V = Ker\ p_1(\mathcal{O}) + Ker\ p_2(\mathcal{O})$.

Next if $v \in Ker\ p_1(\mathcal{O}) \cap Ker\ p_2(\mathcal{O})$, then from the identity above we get $v = q_1(\mathcal{O})p_1(\mathcal{O})v + q_2(\mathcal{O})p_2(\mathcal{O})v$. But the right-hand side of the last equation is just 0 contravening to the left-hand side nonzero v. So $V = Ker\ p_1(\mathcal{O}) \oplus Ker\ p_2(\mathcal{O})$ holds.

Corollary(10.1.4). *Suppose that $p(\mathbb{X}) \in F[\mathbb{X}]$ is a nonconstant polynomial and $p(\mathbb{X}) = p_1(\mathbb{X}) \cdots p_\ell(\mathbb{X})$ for some relatively prime polynomials p_i in $F[\mathbb{X}]$. Let $p(\mathcal{O}) = 0$ for an operator $\mathcal{O}$ of V over F. We then have $V = Ker\ p_1(\mathcal{O}) \oplus \cdots \oplus Ker\ p_\ell(\mathcal{O})$, where each direct summand is $\mathcal{O}$-invariant.*

Proof. By induction on ℓ and by using the proposition, we can easily prove it. Indeed $V = Ker\ p_1(\mathcal{O}) \oplus Ker\ p_2(\mathcal{O}) \cdots p_\ell(\mathcal{O})$ holds, whence applying the induction hypothesis to $Ker\ p_2(\mathcal{O})$

$\cdots p_\ell(\mathcal{O})$ yields

$$Ker\ p_2(\mathcal{O})\cdots p_\ell(\mathcal{O}) = Ker\ p_2(\mathcal{O}) \oplus \cdots \oplus Ker\ p_\ell(\mathcal{O}).$$

But we still have to show that $Ker\ p_i(\mathcal{O})$ in V coincides with $Ker\ p_i(\mathcal{O})$ in $Ker\ p_2(\mathcal{O})\cdots p_\ell(\mathcal{O})$ for $i = 2, \cdots, \ell$. Now if $v \in Ker\ p_i(\mathcal{O}), i = 2, \cdots, \ell$, then $p_2(\mathcal{O})\cdots p_\ell(\mathcal{O})v = 0$. Hence $v \in Ker\ p_2(\mathcal{O})\cdots p_\ell(\mathcal{O})$, thereby proving our proposition.

Example(10.1.5). Let $p(\mathbb{X}) \in \mathbb{C}[\mathbb{X}]$ have a factorization like $p(\mathbb{X}) = (\mathbb{X} - c_1)^{i_1}\cdots(\mathbb{X} - c_\ell)^{i_\ell}$ with $c_i's$ mutually distinct. Let $p(\mathcal{O}) = 0$ for an operator $\mathcal{O}$ of a $\mathbb{C}$-vector space V. We then have $V = Ker(\mathcal{O} - c_1 I)^{i_1} \oplus \cdots \oplus Ker(\mathcal{O} - c_\ell I)^{i_\ell}$ by virtue of corollary 10.1.4.

Exercises of § 10-1

[1] Verify $Ker\ f \cdot p(\mathcal{O}) = Ker\ p(\mathcal{O})$ $\forall f(\neq 0) \in F$, $\forall p(\mathbb{X}) \in F[\mathbb{X}]$, and for any operator $\mathcal{O}$ of an F-vector space V.

[2] In proposition 10.1.3, it holds that $Im\ p_2(\mathcal{O}) = Ker\ p_1(\mathcal{O})$. Prove this.

[3] Assume the dimension of each eigenspace of $\mathcal{O} \leq 1$. If $p(\mathcal{O}) = (\mathcal{O} - f_1 I) \cdots (\mathcal{O} - f_\ell I) = 0$ for distinct $f_i \in F$ and an operator $\mathcal{O}$ of an F-vector space V with $dim V = \ell$. Show then V has a basis consisting of eigenvectors of $\mathcal{O}$ with corresponding eigenvalues $f_1, \cdots, f_\ell$ in F.

§ 10-2. $\widetilde{S}$-invariant space

Let V be a vector space over F and let $\widetilde{S}$ be a set of operators of V. If an operator commutes with each of $\widetilde{S}$, some interesting consequences are then attained.

Definition(10.2.1). In the notation of above, by an *irreducible* $\widetilde{S}$-space or a *simple* $\widetilde{S}$-space V we mean a nonzero vector space V over F whose $\widetilde{S}$-invariant subspaces are only (0) and V itself.

Proposition(10.2.2). *With $\widetilde{S}$ as above let $\mathcal{O} : V \longrightarrow V$ be an operator satisfying $\mathcal{OP} = \mathcal{PO}\ \forall \mathcal{P} \in \widetilde{S}$. $Im\mathcal{O}$ and $Ker\mathcal{O}$ are then $\widetilde{S}$-invariant subspaces of V.*

Proof. Evident because $\mathcal{OP} = \mathcal{PO}$ holds.

Now either one of the following propositions is called "Schur lemma".

Proposition(10.2.3). *As above, suppose that $\widetilde{S}$ is a set of operators of a vector space V over F and that an operator $\mathcal{O} : V \longrightarrow V$ commutes with any of $\widetilde{S}$. In addition let V be an irreducible $\widetilde{S}$-space. We then have either $\mathcal{O}$ is trivial or invertible.*

Proof. Suppose that $\mathcal{O}$ is not trivial; then by proposition 10.2.2 we see that $Im\mathcal{O}$ is $\widetilde{S}$-invariant, so that $Im\mathcal{O} = V$. On the other hand $Ker\mathcal{O}$ is also $\widetilde{S}$-invariant, so that $Ker\mathcal{O} = \{0\}$. Consequently $\mathcal{O}$ is injective by proposition 4.1.12, whence $\mathcal{O}$ is invertible.

Proposition(10.2.4). *Suppose that V is a finite dimensional vector space over an algebraically closed field F and that V is an irreducible $\widetilde{S}$-space with $\widetilde{S}$ a set of operators of V. In addition let an operator $\mathcal{O}$ of V commute with any operator $\mathcal{P}$ of $\widetilde{S}$. We then have $\mathcal{O} = fI$ for some $f \in F$.*

Proof. We designate $J := \{p(\mathbb{X}) \in F[\mathbb{X}] | p(\mathcal{O}) = 0\}$ which is obviously an ideal of $F[\mathbb{X}]$. Since $p_{\mathcal{O}}(\mathcal{O}) = 0$, $J \neq \{0\}$ in the case of $\mathcal{O} \neq 0$. Let g be a generator of J from the standpoint of proposition 9.4.2. We claim $g(\mathbb{X})$ is irreducible, namely $g(\mathbb{X}) = \mathbb{X} - f$ for some $f \in F$ if we select the leading coefficient 1 of g.

If we deny this claim, then we have $g = h_1 h_2$ for some non-constant polynomials $h_1, h_2 \in F[\mathbb{X}]$. Since $h_1(\mathcal{O}), h_2(\mathcal{O}) \neq 0$ because of degrees $< deg\ g(\mathbb{X})$, we see that $h_1(\mathcal{O}), h_2(\mathcal{O})$ are invertible in view of the proposition 10.2.3. We thus have an invertible operator $h_1(\mathcal{O}) \circ h_2(\mathcal{O}) = g(\mathcal{O})$, a contradiction showing that $0 = g(\mathcal{O}) = \mathcal{O} - fI$.

Remark(10.2.5). Note that the finite dimensionality of F-vector space V is necessary in the above proposition. Otherwise we can't assure the ideal J of $F[\mathbb{X}]$ which is nonzero.

Exercises of § 10-2

[1] Let V be an F-vector space and let $dim_F V < \infty$, where F is algebraically closed. In the notation of proposition 10.2.3, show that V has a basis consisting of eigenvectors with a single eigenvalue.

[2] Suppose that $\widetilde{S}$ consists of the matrix $\begin{pmatrix} 1 & r \\ 0 & 1 \end{pmatrix}$ which may be viewed as an operator of $\mathbb{R}^2$, where $\forall r \in \mathbb{R}$. Find out all $\widetilde{S}$-invariant subspaces of $\mathbb{R}^2$.

[3] Use the Zorn's lemma or equivalently the well-ordering theorem to show that any F-vector space V has a basis and so does any $\widetilde{S}$-invariant subspace of V as above.

[4] Determine all $\widetilde{S}$-invariant subspaces of V, where $\widetilde{S} = LM(V, V)$ and V is any F-vector space. (Hint : use [3] above).

§ 10-3. Cyclic vector space

In the next section we shall see that a finite dimensional vector space over an algebraically closed field is decomposed into cyclic subspaces with respect to any operator. For this important result we should be prepared for some lemmas relevant to linear independency.

Definition(10.3.1). Suppose that V is a vector space over an algebraically closed field F. Let $\mathcal{O} : V \longrightarrow V$ be an operator of V. For $v \neq 0 \in V$ and $f \in F$, v is said to be $(\mathcal{O} - fI)$-*cyclic* if $(\mathcal{O} - fI)^s v = 0$ but $(\mathcal{O} - fI)^{s-1} v \neq 0$ for some number $s \in \mathbb{N}$.

In this case we shall say that v *has the period* s with respect to $\mathcal{O} - fI$. In case that v is $(\mathcal{O} - fI)$-cyclic and $\{v, \mathcal{O}v, \cdots, \mathcal{O}^{s-1}v\}$ generates the F-vector space V, we shall say that V is *cyclic.*

Lemma(10.3.2). *The subspace of V generated by $\{v, \mathcal{O}v, \cdots, \mathcal{O}^{s-1}v\}$ is equal to that of V generated by $\{v, (\mathcal{O} - fI)v, \cdots, (\mathcal{O} - fI)^{s-1}v\}$ in the notation as above.*

Proof. $\forall\, p(\mathbb{X}) \in F[\mathbb{X}]$ may be expressed as a power series in $\mathbb{X} - f$ $\forall f \in F$ and vice versa. So our claim is evident.

Lemma(10.3.3). *If v is $(\mathcal{O} - fI)$-cyclic with period s, then $\{(\mathcal{O} - fI)^{s-1}v, \cdots, (\mathcal{O} - fI)v, v\}$ is a linearly independent set of V.*

Proof. Assume that we have a linear dependence equation $p(\mathcal{O} - fI)v = 0$ for some nonzero polynomial $p(\mathbb{X})$ of degree $r \leq s - 1$.

But we have $(\mathcal{O} - fI)^s v = 0$ by the condition as well. Putting $q(\mathbb{X}) = \mathbb{X}^s$ and $g(\mathbb{X})$:=g.c.d.(p, q) gives rise to $g(\mathbb{X}) = \widetilde{p}p + \widetilde{q}q$ for some $\widetilde{p}, \widetilde{q} \in F[\mathbb{X}]$ by virtue of proposition 9.4.3. Replacing $\mathbb{X}$ by $\mathcal{O} - fI$, however, yields $g(\mathcal{O} - fI) = \widetilde{p}(\mathcal{O} - fI)p(\mathcal{O} - fI) + \widetilde{q}(\mathcal{O} - fI)q(\mathcal{O} - fI)$, so that $g(\mathcal{O} - fI)v = 0$ necessarily happens. In the mean time it follows that $g(\mathbb{X}) = \mathbb{X}^m$ with $m \leq s-1$ since $g(\mathbb{X})$:=g.c.d. of (p, q), which contravenes to the period s of v. So $(\mathcal{O} - fI)^{s-1}v, \cdots, v$ are linearly independent.

Proposition(10.3.4). *Let V be a cyclic vector space over F as above with respect to an operator $\mathcal{O}$ of V. Then for some basis $\mathcal{B}$ of V*

$$A_{\mathcal{B}}^{\mathcal{B}}(\mathcal{O}) = \begin{pmatrix} f & 1 & 0 & \cdots & 0 \\ 0 & \ddots & \ddots & \ddots & \vdots \\ \vdots & \ddots & \ddots & \ddots & 0 \\ \vdots & & \ddots & \ddots & 1 \\ 0 & \cdots & \cdots & 0 & f \end{pmatrix}$$

holds.

Proof. By virtue of the above two lemmata we see that the set $\mathcal{B} := \{(\mathcal{O} - fI)^{s-1}v, \cdots, (\mathcal{O} - fI)v, v\}$ is a basis of V. Now $\forall$ basis element $(\mathcal{O} - fI)^r v$ with $r \leq s-1$, we have $\mathcal{O} \cdot (\mathcal{O} - fI)^r v = (\mathcal{O} - fI)^{r+1}v + f(\mathcal{O} - fI)^r v$ completing the proof.

Exercises of § 10-3

[1] Let V be a finite dimensional vector space over a field F and $\mathcal{O}$ an operator of V. If $\exists n \in \mathbb{N}$ such that $\mathcal{O}^n = 0$, then we call $\mathcal{O}$ nilpotent. Show that if V is cyclic with respect to $\mathcal{O}$, then $\mathcal{O} = D+N$ for some diagonal operator D and some nilpotent operator N such that $DN = ND$.

[2] If $v \in V$ is a nonzero $(\mathcal{O} - fI)$-cyclic vector of period $s \in \mathbb{N}$, show that if V is cyclic, then V is essentially no cyclic with respect to $(\mathcal{O} - \widetilde{f}I)$-cyclic vector with $\widetilde{f}(\neq f)$ in F.

[3] Let V be a finite dimensional vector space over an algebraically closed field F. Suppose that ε is an eigenvalue of an operator $\mathcal{O}$. Show that V contains at least one $(\mathcal{O} - \varepsilon I)$-cyclic vector.

[4] If V is any finite dimensional cyclic vector space over any field F, show then that any nonzero $v \in V$ is a cyclic vector with respect to an operator.

§ 10-4. Jordan basis and its associated matrix

Let V be a finite dimensional vector space over an algebraically closed field F and $\mathcal{O}$ an operator of V. If V is decomposed into a direct sum of $\mathcal{O}$-invariant cyclic subspaces V_i of V like $V = V_1 \oplus \cdots \oplus V_k$, then we may exhibit a basis of V using each convenient basis like that of proposition 10.3.3.

Definition(10.4.1). In the notation as above, the basis $\mathcal{B}_i$ of each cyclic subspace V_i as in proposition 10.3.3 is called the *Jordan basis* of V_i relative to $\mathcal{O}$. Furthermore the Jordan basis of V relative to $\mathcal{O}$ is just $\mathcal{B} := \mathcal{B}_1 \cup \mathcal{B}_2 \cup \cdots \cup \mathcal{B}_k$ with respect to which $A_{\mathcal{B}}^{\mathcal{B}}(\mathcal{O})$ looks like

$$\begin{pmatrix} \boxed{\begin{matrix} \varepsilon_1 & 1 & & \\ & \ddots & \ddots & \\ & & \ddots & 1 \\ & & & \varepsilon_1 \end{matrix}} & & & \\ & \ddots & & \\ & & \ddots & \\ & & & \boxed{\begin{matrix} \varepsilon_k & 1 & & \\ & \ddots & \ddots & \\ & & \ddots & 1 \\ & & & \varepsilon_k \end{matrix}} \end{pmatrix}.$$

where each block has an eigenvalue ε_i on the diagonal while each superdiagonal elements in each block are only 1, if any, and everywhere else is zero.

It is notable that the above decomposition is possible in the above setting, which is explicitly understood by the final goal.

Jordan decomposition theorem(10.4.2). *Suppose that V is a finite dimensional vector space over an algebraically closed field F. Let $\mathcal{O}$ be any operator of V. We then have $\mathcal{O}$-invariant cyclic subspaces V_i such that $V = V_1 \oplus \cdots \oplus V_\ell$.*

Proof. By virtue of corollary 10.1.4 and definition 9.3.5, we may assume that $(\mathcal{O}-\varepsilon I)^s V = 0$ but $(\mathcal{O}-\varepsilon I)^{s-1} V \neq 0$ for some $s \in \mathbb{N}$ and $\varepsilon \in F$. Since $dim\ I_m(\mathcal{O}-\varepsilon I)+dim\ Ker(\mathcal{O}-\varepsilon I) = dim\ V$, we see $dim\ I_m(\mathcal{O}-\varepsilon I) < dim\ V$. So applying induction hypothesis to the subspace $(\mathcal{O}-\varepsilon I)V$ of V gives rise to the decomposition of $(\mathcal{O}-\varepsilon I)V$ into a direct sum of $\mathcal{O}$-invariant cyclic subspaces $\widetilde{V}_i$. For instance let $\widetilde{V}_1 \oplus \cdots \oplus \widetilde{V}_k = (\mathcal{O}-\varepsilon I)V$, where each $\widetilde{V}_i$ has a Jordan basis $\{\widetilde{v_i}, (\mathcal{O}-\varepsilon_i I)\widetilde{v_i}, \cdots, (\mathcal{O}-\varepsilon_i I)^{s_i-1}\widetilde{v_i}\}$ for $i = 1, \cdots, k$. If $(\mathcal{O}-\varepsilon_i I)v_i = \widetilde{v_i}$ for some v_i of V, then each v_i is also $(\mathcal{O}-\varepsilon_i I)$-cyclic.

Let V_i be the subspace of V generated by $v_i, (\mathcal{O}-\varepsilon_i I)v_i, \cdots,$ $(\mathcal{O}-\varepsilon_i I)^{s_i} v_i$ for $i = 1, \cdots, k$ which are linearly independent. Let $\widetilde{V}$ be the subspace of V expressed as $\widetilde{V} = V_1 + \cdots + V_k$ which we claim this sum is a direct sum.

For this proof we first note that any vector in V_i is of the form $p_i(\mathcal{O}-\varepsilon_i I)v_i$ for some polynomial $p_i(\mathbb{X})$ of degree$\leq s_i$. Now assume that $\exists$ a linear dependence relation $\sum_{i=1}^{k} p_i(\mathcal{O}-\varepsilon I)v_i = 0$. Multiplying $(\mathcal{O} - \varepsilon I)$ on both sides of this equation yields $\sum_{i=1}^{k} p_i(\mathcal{O}-\varepsilon I)\widetilde{v_i} = 0$, so that $p_i(\mathcal{O}-\varepsilon_i I)\widetilde{v_i} = 0$ for $i = 1, \cdots, k$ because $(\mathcal{O} - \varepsilon I)V = \widetilde{V_1} \oplus \cdots \oplus \widetilde{V_k}$.

So $\mathbb{X}^{s_i}|p_i(\mathbb{X})$ with $s_i \geq 1$ gives $\mathbb{X}|p_i(\mathbb{X})$ in particular, and thus we may put $p_i(\mathbb{X}) = \mathbb{X} \cdot \widetilde{p_i}(\mathbb{X})$ for some $\widetilde{p_i}(\mathbb{X}) \in F[\mathbb{X}]$. But then $0 = \sum_{i=1}^{k} p_i(\mathcal{O} - \varepsilon I)v_i = \sum_{i=1}^{k} \widetilde{p_i}(\mathcal{O} - \varepsilon I)\{(\mathcal{O} - \varepsilon I)v_i\} = \sum_{i=1}^{k} \widetilde{p_i}(\mathcal{O} - \varepsilon I)\widetilde{v_i}$ gives rise to the fact $\widetilde{p_i}(\mathcal{O} - \varepsilon I)\widetilde{v_i} = 0 = \widetilde{p_i}(\mathcal{O} - \varepsilon I)(\mathcal{O} - \varepsilon I)v_i = p_i(\mathcal{O} - \varepsilon I)v_i$. Hence $\widetilde{V}$ is a direct sum as claimed.

On the other hand since any vector of $(\mathcal{O} - \varepsilon I)V$ is of the style $\sum_{i=1}^{k} p_i(\mathcal{O} - \varepsilon I)\widetilde{v_i}$ for any $p_i(\mathbb{X}) \in F[\mathbb{X}]$, we easily notice that $Im(\mathcal{O} - \varepsilon I)$ in V $=(\mathcal{O} - \varepsilon I)\widetilde{V}$. We thus understand $V = Ker(\mathcal{O} - \varepsilon I) + \widetilde{V}$ because $\forall v \in V, (\mathcal{O} - \varepsilon I)v = (\mathcal{O} - \varepsilon I)w$ for some $w \in \widetilde{V}$ and $(\mathcal{O}-\varepsilon I)(v-w) = 0$ gives $v - w \in Ker(\mathcal{O}-\varepsilon I)$. $Ker(\mathcal{O}-\varepsilon I)+\widetilde{V}$ may not be direct sum; but after choosing some 1 dimensional cyclic $\mathcal{O}$-invariant subspaces $K_i's$ we may make $\widetilde{V} \oplus K_1 \oplus \cdots \oplus K_m = V$, namely $V_1 \oplus \cdots \oplus V_k \oplus K_1 \oplus \cdots \oplus K_m = V$. After all V can be expressed in the form $V_1 \oplus \cdots \oplus V_\ell = V$ with each V_i $\mathcal{O}$-invariant cyclic in V.

Remark(10.4.3). F being an algebraically closed field is crucial in the Jordan decomposition theorem. In particular $\mathbb{C}$ is algebraically closed. We shall exhibit some references suggesting its proof in the appendix. But note that all of them make use of non-algebraic ideas.

Exercises of § 10-4

[1] Verify that the elementary divisors of the characteristic polynomial of a real symmetric matrix or a hermitian matrix are all linear polynomials.

[2] Show that the elementary divisors of the characteristic polynomial of a complex unitary operator are all linear polynomials.

[3] Let V be a finite dimensional vector space over any algebraically closed field F. Let $\mathcal{O} : V \longrightarrow V$ be an operator.

(i) Prove that $\mathcal{O}$ may be expressed in the form $\mathcal{O} = D + N$ for some diagonal operator D and for some nilpotent operator N which commute with each other.

(ii) In addition, let V be a cyclic vector space. Verify that there exists only one 1-dimensional eigenspace of $\mathcal{O}$.

[4] With assumptions as in [3] before (i),(ii), decompose $V = F^4$ into the direct sum of cyclic subspaces of V if $\mathcal{O}$ is given in the following form:

(i) $\begin{pmatrix} 1 & 0 & 0 & 0 \\ 0 & 1 & 0 & 0 \\ 0 & 0 & 2 & 1 \\ 0 & 0 & 0 & 2 \end{pmatrix}$ (ii) $\begin{pmatrix} 2 & 1 & 0 & 0 \\ 0 & 2 & 0 & 0 \\ 0 & 0 & -2 & 1 \\ 0 & 0 & 0 & -2 \end{pmatrix}$

(iii) $\begin{pmatrix} 2 & 1 & 0 & 0 \\ 0 & 2 & 0 & 0 \\ 0 & 0 & 2 & 0 \\ 0 & 0 & 0 & -2 \end{pmatrix}$ (iv) $\begin{pmatrix} -4 & 1 & 0 & 0 \\ 0 & -4 & 1 & 0 \\ 0 & 0 & -4 & 1 \\ 0 & 0 & 0 & -4 \end{pmatrix}$

(v) $$\begin{pmatrix} -4 & 0 & 0 & 0 \\ 0 & -3 & 0 & 0 \\ 0 & 0 & -2 & 0 \\ 0 & 0 & 0 & -1 \end{pmatrix}$$

[5] For a finite dimensional vector space V over a field F and for an operator $\mathcal{O}$ of V, by a fan of $\mathcal{O}$ we mean a sequence of $\mathcal{O}$-invariant subspaces $\{S_1, \cdots, S_\ell\}$ with $V = S_\ell$ such that $dim S_i = i$ for $i = 1, \cdots, \ell$ and S_i is included in S_{i+1} for $i = 1, \cdots, \ell - 1$.

If $\{v_1, \cdots, v_i\}$ is a basis of S_i, we call it a fan basis of S_i. In this case show that the matrix associated with $\mathcal{O}$ with respect to a fan basis of $\mathcal{O}$ becomes upper triangular. Also show that a Jordan basis is a fan basis.

APPENDIX

For the proof of the field $\mathbb{C}$ being algebraically closed, see the following citations:

[F] John B. Fraleigh, A First Course in Abstract Algebra, Theorem 38.7(Fundamental theorem of algebra), 3rd edition, Addison-Wesley, 1982.

[H] Thomas W.Hungerford, ALGEBRA, Theorem 3.19(The Fundamental theorem of Algebra), New York: Holt, Rinehart and Winston, 1974.

[L] Serge Lang, Linear Algebra, Appendix, 3rd edition, Springer Verlag, 1987.

REFERENCES

[BM] G.Birkhoff and S.MacLane, Algebra, the Macmillan company, 1967.

[F] John B. Fraleigh, A First Course in Abstract Algebra, 3rd edition, Addison-Wesley.

[Ha] P.R.Halmos, Naive set theory, Van Nostrand Reinhold company.

[Hu] Thomas W.Hungerford, ALGEBRA, New York: Rinehart and Winston, 1974.

[Ki] Y.G.Kim, Logic and Set theory(in Korean), Kyung Moon publishers, 2001.

[La-1] Serge Lang, Linear Algebra, 3rd edition, Springer.

[La-2] ———, Algebra, Addison-Wesley, 1984.

[Li] S.Lipschutz, Set theory and related topics, Schaum's outline series, Mcgraw-Hill book company.

SUBJECT INDEX

SKETCH OF ANSWERS TO EXERCISES

Exercises of § 1-1 (page 3)

[1] We only show that $\mathbb{C}$ satisfies definition 1.1.1, the other cases being similar with $\mathbb{Q} \subset \mathbb{R} \subset \mathbb{C}$.

(i) $\forall a, b \in \mathbb{C}, a + b = b + a \in \mathbb{C}$ and $a \cdot b = b \cdot a \in \mathbb{C}$.

(ii) $\forall a, b, c \in \mathbb{C}, (a+b)+c = a+(b+c)$ and $(ab)c = a(bc)$.

(iii) $\forall a, b, c \in \mathbb{C}, (a + b)c = ac + bc$.

(iv) $\exists 0 = 0+0i, 1 = 1+0i$ satisfying that $\forall a \in \mathbb{C}, a+0 = a$ and $a \cdot 1 = a$.

(v) $\forall a \in \mathbb{C}, \exists -a \in \mathbb{C}$ such that $a + (-a) = 0$.
$\forall a \in \mathbb{C}^{\times}, \exists\, a^{-1} \in \mathbb{C}$ such that $a \cdot a^{-1} = 1$.

[2] Suppose that $0, 0'$ satisfy (iv) of definition 1.1.1; we then have that $0' = 0' + 0 = 0 + 0' = 0$. Similarly suppose that $1, 1'$ satisfy the latter part of (iv) of definition 1.1.1; we then have $1' = 1' \cdot 1 = 1 \cdot 1' = 1$.

[3] We usually require that $0, 1$ are different.

[4] $\mathbb{Z}_2 = \{0, 1\}$ with the following addition and multiplica-

Typeset by $\mathcal{A}\mathcal{M}\mathcal{S}$-TEX

tion table:

+	0	1		×	0	1
0	0	1		0	0	0
1	1	0		1	0	1

[5] If we assume that 0^{-1} exists and $0^{-1} = x \in F$, we then $1 = 0 \cdot 0^{-1} = 0 \cdot x = 0$, which is a contradiction.

Exercises of § 1-2 (page 7)

[1] $\forall v \in V, 0v = (0+0)v = 0v + 0v$, so that $-0v + 0v = -0v + (0v + 0v) = (-0v + 0v) + 0v$ holds. Hence $0v = \mathbb{O}$. Similarly $\forall f \in F$, we have $f\mathbb{O} = f(\mathbb{O} + \mathbb{O}) = f\mathbb{O} + f\mathbb{O}$, so that $f\mathbb{O} = \mathbb{O}$ is obtained subtracting $f\mathbb{O}$ from both sides.

[2] If $v = v'$, then $fv = f'v' = f'v$, so that $(f - f')v = \mathbb{O}$. If $f \neq f'$ in this equation, we then have $(f - f')^{-1}(f - f')v = (f - f')^{-1}\mathbb{O}$, which implies $1v = v = \mathbb{O}$, a contradiction. Hence $f = f'$.

Similarly if $f = f'$, then $fv = f'v' = fv'$, so that $f(v - v') = \mathbb{O}$. Since f can't be zero, we obtain $f^{-1}f(v - v') = f^{-1}\mathbb{O} = \mathbb{O}$. Hence $1(v - v') = v - v' = \mathbb{O}$, so that $v = v'$ is obtained.

[3] We have only to show that $\forall\ f, f' \in F, \forall w, w' \in S, fw + f'w' \in S$. However we compute $v(fw + f'w') = v(fw) + v(f'w') = f(vw) + f'(vw') = f0 + f'0 = 0 + 0 = 0$. Hence we get $fw + f'w' \in S$ as desired.

[4]

(i) $\forall \widetilde{f_1}, \widetilde{f_2} \in V,\ \widetilde{f_1} + \widetilde{f_2} = \widetilde{f_2} + \widetilde{f_1}$.

(ii) $\forall f \in F,\ \forall \widetilde{f_1} \in V,\ f\widetilde{f} \in V$ is obtained.

(iii) $\widetilde{f} \in V,\ 1 \cdot \widetilde{f} = \widetilde{f}$ holds.

(iv) $\forall \widetilde{f_1}, \widetilde{f_2}, \widetilde{f_3} \in V$, we have $\widetilde{f_1}+(\widetilde{f_2}+\widetilde{f_3}) = (\widetilde{f_1}+\widetilde{f_2})+\widetilde{f_3}$.

(v) We define $\mathbb{O} : E \to F$ by $\mathbb{O}(e) = 0\ \forall e \in E$. We then have $\forall \widetilde{f} \in V, \widetilde{f} + \mathbb{O} = \mathbb{O} + \widetilde{f} = \widetilde{f}$.

(vi) $\forall \widetilde{f} \in V$, we define $-\widetilde{f} := (-1)\widetilde{f}$. We then have $\widetilde{f} + (-\widetilde{f}) = (-\widetilde{f}) + \widetilde{f} = \mathbb{O}$ easily.

(vii) $\forall f \in F, \forall \widetilde{f_1}, \widetilde{f_2} \in V$, we get $f(\widetilde{f_1} + \widetilde{f_2}) = f\widetilde{f_1} + f\widetilde{f_2}$.

(viii) $\forall f_1, f_2 \in F, \forall \widetilde{f} \in V$, we get $(f_1 + f_2)\widetilde{f} = f_1\widetilde{f} + f_2\widetilde{f}$.

(ix) $\forall f_1, f_2 \in F, \forall \widetilde{f} \in V$, we have $(f_1 f_2)\widetilde{f} = f_1(f_2\widetilde{f})$.

$\therefore V$ forms a vector space over F.

[5] Since $x = x + 0\sqrt{2}\ \ \forall x \in \mathbb{Q}$, we obviously have $\mathbb{Q} \subset V$. Now we check if V becomes a field:

(i) $\forall a, b \in V,\ a + b = b + a \in V$ and $ab = ba \in V$.

(ii) $\forall a, b, c \in V, (a+b)+c = a+(b+c)$ and $(ab)c = a(bc)$.

(iii) $\forall a, b, c \in V$, we have $(a + b)c = ac + bc$.

(iv) $\exists 0 = 0 + 0\sqrt{2}$ and $1 = 1 + 0\sqrt{2}$ such that $\forall a = x + y\sqrt{2} \in V, a + 0 = a$ and $a \cdot 1 = a$.

(v) $\forall a = x+y\sqrt{2}, \exists -a = -x-y\sqrt{2}$ such that $a+(-a) = 0$. $\forall a \in V - \{0\}, \exists a^{-1} \in V$ such that $a \cdot a^{-1} = 1$ noting that for $a = x + y\sqrt{2},\ a^{-1} = \frac{1}{x+y\sqrt{2}} = \frac{x-y\sqrt{2}}{(x+y\sqrt{2})(x-y\sqrt{2})} = \frac{x-y\sqrt{2}}{x^2-2y^2} = \frac{x}{x^2-2y^2} - \frac{y}{x^2-2y^2}\sqrt{2}$.

Next we show that V becomes a vector space over

$\mathbb{Q}$ and also over V itself.

(i) $\forall v_1, v_2 \in V, v_1 = x_1 + y_1\sqrt{2}, v_2 = x_2 + y_2\sqrt{2}$ for some $x_i, y_i \in \mathbb{Q}$. Now $v_1 + v_2 = (x_1 + x_2) + (y_1 + y_2)\sqrt{2} = (x_2 + x_1) + (y_2 + y_1)\sqrt{2} = v_2 + v_1 \in V$.

(ii) $\forall q \in \mathbb{Q}, \forall v = x + y\sqrt{2} \in V, fv = f(x + y\sqrt{2}) = fx + fy\sqrt{2} \in V$.

(iii) $\forall v = x + y\sqrt{2}, 1v = 1(x + y\sqrt{2}) = 1x + 1y\sqrt{2} = x + y\sqrt{2} = v$.

(iv) $\forall v_i = x_i + y_i\sqrt{2} \in V$ for $i = 1, 2, 3$, we have $v_1 + (v_2 + v_3) = (x_1 + y_1\sqrt{2}) + (x_2 + y_2\sqrt{2} + x_3 + y_3\sqrt{2}) = \{x_1 + (x_2 + x_3)\} + \{y_1 + (y_2 + y_3)\}\sqrt{2} = \{(x_1 + x_2) + x_3\} + \{(y_1 + y_2) + y_3\}\sqrt{2} = (v_1 + v_2) + v_3$.

(v) $\exists \mathbb{O} = 0 + 0\sqrt{2}$ such that $\forall v = x + \sqrt{2}y \in V, v + \mathbb{O} = \mathbb{O} + v = v$.

(vi) $\exists -v = -x - y\sqrt{2},\ \forall v = x + y\sqrt{2} \in V$ such that $v + (-v) = -v + v = \mathbb{O}$.

(vii) $\forall q \in \mathbb{Q}, \forall v_1, v_2 \in V,\ q(v_1 + v_2) = qv_1 + qv_2$.

(viii) $\forall v \in V, \forall q_1, q_2 \in \mathbb{Q},\ (q_1 + q_2)v = q_1 v + q_2 v$.

(ix) $\forall v \in V, \forall q_1, q_2 \in \mathbb{Q},\ (q_1 q_2)v = q_1(q_2 v)$.

Hence V is a vector space over $\mathbb{Q}$. Similarly we may show that V is a vector space over itself by making use of properties of $\mathbb{R}$.

[6] Let F be any fixed field. We check axioms:

(i) $\forall f_1, f_2 \in F,\ f_1 + f_2 = f_2 + f_1 \in F$.

(ii) $\forall f \in F, \forall v \in F,\ fv \in F$.

(iii) $\forall f \in F,\ 1 \cdot f = f$.

(iv) $\forall f_1, f_2, f_3 \in F$, we have $f_1+(f_2+f_3) = (f_1+f_2)+f_3$.

(v) $\exists 0 \in F$ such that $\forall f \in F,\ f + 0 = 0 + f = f$.

(vi) $\forall f \in F,\ f + (-f) = -f + f = 0$.

(vii) $\forall f \in F, \forall v_1, v_2 \in F,\ f(v_1 + v_2) = fv_1 + fv_2$.

(viii) $\forall f_1, f_2 \in F, \forall v \in F,\ (f_1 + f_2)v = f_1 v + f_2 v$.

(ix) $\forall f_1, f_2 \in F, \forall v \in F,\ (f_1 f_2)v = f_1(f_2 v)$.

So F is a vector space over itself.

Exercises of § 1-3 (page 11)

[1] If we let $r_1 e^x + r_2 e^{4x} = 0$ for some $r_1, r_2 \in \mathbb{R}$, we then have $r_1 e^x + 4r_2 e^{4x} = 0$ by differentiating the given equation. Subtracting one equation from the other, we obtain $r_1 = r_2 = 0$. Hence e^x, e^{4x} are linearly independent.

[2] $V_S = \{r_1 e^x + r_2 e^{4x} \mid \forall r_1, r_2 \in \mathbb{R}\}$.

[3] Since $e^x, 2e^x$ are linearly dependent as is shown by the equation $2e^x - (2e^x) = 0$, $\{e^x, 2e^x, e^{4x}\}$ cannot be a basis of V_S of [2].

[4] $r_1(1,1) + r_2(0,2) = 0 \Rightarrow r_1 = 0$ and $r_2 = 0$.

So $(1,1), (0,2)$ are linearly independent over $\mathbb{R}$. Next $\forall (r_1, r_2) \in \mathbb{R}^2$, we put $(r_1, r_2) = x_1(1,1) + x_2(0,2)$ and get $x_1 = r_1, x_2 = \frac{1}{2}(r_2 - r_1)$.

So$\{(1,1),(0,2)\}$ generates $\mathbb{R}^2$. Hence $\{(1,1),(0,2)\}$ becomes a basis of $\mathbb{R}^2$.

[5] $(2,4) = 2(1,0) + 4(0,1) = 2(1,1) + 1(0,2)$, so that coordinates of $(2,4)$ are $(2,4), (2,1)$ respectively with respect to bases $\{(1,0),(0,1)\}, \{(1,1),(0,2)\}$ respectively.

[6] V is infinite-dimensional. More precisely $Dim_{\mathbb{R}}(V)$ becomes the cardinality of a denumerable set considering Fourier series and $\mathbb{P}^2(\mathbb{R})$ which is the 2-dimensional projective space over $\mathbb{R}$.

[7] For V as in [1],$\{e^x, e^{4x}\} \subset \{e^x, 2e^x, e^{4x}\}$ is a maximal subset of linearly independent vectors.

So $\{e^x, e^{4x}\}$ is a basis of V_S over $\mathbb{R}$. More explicitly, $r_1e^x + r_2e^{4x} = 0$ for some $r_1, r_2 \in \mathbb{R}$ yields $r_1e^x + 4r_2e^{4x} = 0$ by differentiating both sides. Subtracting one equation from the other, we have $r_2 = r_1 = 0$. So we finally get $2 = Dim_{\mathbb{R}}V_S$.

[8] For infinite dimension, see theorem 2.6 chapter 4 [Hu]. It says that any two bases of an infinite dimensional vector spaces have the same cardinality.

For finite dimension, we suppose first that we have bases $\{v_1, \cdots, v_\ell\}, \{w_1, \cdots, w_k\}$ of the same vector space V over F. But then it follows that $\ell \leq k$ and $\ell \geq k$ by virtue of proposition 1.3.10. So we have $\ell = k$ which justifies the definition of $Dim_F V$.

Exercises of § 1-4 (page 15)

[1] We have $S_1 \times \cdots \times S_\ell = (S_1, 0, \cdots, 0) \oplus \cdots \oplus (0, \cdots, 0, S_\ell)$.

Since $\sharp(0, \cdots, 0, S_i, 0, \cdots, 0) = \sharp(S_i)$, we have a bijection $S_1 \times \cdots \times S_\ell \longrightarrow S_1 \oplus \cdots \oplus S_\ell$ defined by $(x_1, \cdots, x_\ell) \mapsto x_1 + \cdots + x_\ell$.

[2] $\forall (r_1, r_2) \in \mathbb{R}^2$, let$(r_1, r_2) = x_1(1, -1) + x_2(1, 1)$; we then have $\left\{ \begin{array}{c} x_1 + x_2 = r_1 \\ -x_1 + x_2 = r_2 \end{array} \right\}$. Solving both equations with respect to x_1, x_2, we have $x_1 = \frac{(r_1 - r_2)}{2}, x_2 = \frac{(r_1 + r_2)}{2}$. So $\{(1, -1), (1, 1)\}$ generates $\mathbb{R}^2$,i.e., $\mathbb{R}^2 = \mathbb{R}(1, -1) + \mathbb{R}(1, 1)$.

Now $\mathbb{R}(1, -1) \cap \mathbb{R}(1, 1) = \{\mathbb{O}\}$; otherwise $x(1, -1) = y(1, 1)$ for some x, y (not both zero) in $\mathbb{R}$. But then $\left\{ \begin{array}{c} x = y \\ -x = y \end{array} \right\}$ meaning $x = y = 0$, a contradiction.

$\therefore\ \mathbb{R}^2 = \mathbb{R}(1, -1) \oplus \mathbb{R}(1, 1)$ by virtue of proposition 1.4.4.

[3] Since $\mathbb{Q}(1, -1) \cap \mathbb{Q}(-1, 1) \neq \{\mathbb{O}\}$, the expression $\mathbb{Q}(1, -1) \oplus \mathbb{Q}(-1, 1)$ is wrong according to proposition 1.4.4.

[4] Clearly $\mathbb{C} = \mathbb{R} + \mathbb{R}i$ holds. Furthermore $\mathbb{R} \cap \mathbb{R}i = \{\mathbb{O}\}$, so that $\mathbb{C} = \mathbb{R} \oplus \mathbb{R}i$ holds by virtue of proposition 1.4.4.

[5] $\forall x, x' \in \cap_{i \in I} S_i, \forall f \in F$, we have that $\forall i \in I,\ x, x' \in S_i$. So we see that $\forall i \in I,\ x + x' \in S_i$ and $fx \in S_i$.

$\therefore\ x + x' \in \cap_{i \in I} S_i$ and $fx \in \cap_{i \in I} S_i$.

$\therefore\ \cap_{i \in I} S_i$ becomes a subspace of V.

[6] $V/S = \{v + S \mid \forall v \in V\}$. We define scalar multiplication on V/S by the rule : $\forall f \in F, \forall v + S \in V/S, f(v + S) = fv + S$, which is well defined. Since $(v + S) + (w + S) =$

$(v+w)+S$, it is easy to check that V/S is a vector space over F.

Exercises of § 2-1 (page 19)

[1] $M+M'$ is computable if and only if $(m,n)=(s,t)\in\mathbb{N}^2$.

[2] ${}_tM=(x_{ij})_{n\times m}$ with $x_{ij}=f_{ji}$ and ${}_tM'=(x'_{ij})_{t\times s}$ with $x'_{ij}=f'_{ji}$.

[3] $(f_{ij})_{m\times n}=(f_{jk})_{m\times n}$.

[4] It is easy to show that the set of all $m\times n$ matrices over F forms a vector space V over F under the following operations: For $M=(f_{ij})_{m\times n}$ and $M'=(f'_{ij})_{m\times n}$, we define $M+M':=(f_{ij}+f'_{ij})_{m\times n}$. As regards multiplication by scalar, we define $fM=(ff_{ij})_{m\times n}\ \forall f\in F,\forall M=(f_{ij})_{m\times n}$.

Next let E_{ij} be an elementary matrix whose (i,j)-th entry is just 1 and all others are zero. We then easily see that all $E'_{ij}s$ for $1\le i\le m$ and $1\le j\le n$ form a basis of V over F. So $Dim_F V=m\times n$.

[5] We have already seen in §1-2 that F^ℓ becomes a vector space with componentwise addition and componentwise scalar multiplication prescribed there.

Similarly ${}_t(F^n)=\{{}_t(f_1,\cdots,f_n)\mid\forall f_i\in F\}$ becomes a vector space under similar componentwise addition and componentwise scalar multiplication.

[6] We have seen already that the collection of all $n\times n$

matrices over F forms a vector space over F. So we have only to show that the set S of all $n \times n$ triangular matrices forms a subspace of this vector space:

$\forall (f_{ij}), (f'_{ij}) \in S, \forall f \in F$, we have that $(f_{ij}) + (f'_{ij}) = (f_{ij} + f'_{ij}) \in S$ and that $f(f_{ij}) = (ff_{ij}) \in S$ since $f_{ij} = f'_{ij} = 0$ for $i \gneqq j$ implies $f_{ij} + f'_{ij} = 0$ and $ff_{ij} = 0$ for $i \gneqq j$.

[7] Similar answer to that of [6].

[8] (i) Let $M = (f_{ij})_{m \times m}$ and $N = (g_{jk})_{m \times m}$ for $f_{ij}, g_{jk} \in F$. We then have $MN = (h_{ik})_{m \times m}$ with $h_{ik} = \sum_{j=1}^{m} f_{ij} g_{jk}$, so that $h_{ii} = \sum_{j=1}^{m} f_{ij} g_{ji}$ for $1 \leq i \leq m$.

Similarly for $N = (g_{ij})_{m \times m}, M = (f_{jk})_{m \times m}$, we have $NM = (h'_{ik})_{m \times m}$ with $h'_{ik} = \sum_{j=1}^{m} g_{ij} f_{jk}$, so that $h'_{ii} = \sum_{j=1}^{m} g_{ij} f_{ji}$. Hence we obtain

$$tr(MN) = \sum_{i=1}^{m} h_{ii} = \sum_{i=1}^{m} \sum_{j=1}^{m} f_{ij} g_{ji},$$

while

$$\begin{aligned} tr(NM) &= \sum_{i=1}^{m} h'_{ii} = \sum_{i=1}^{m} \sum_{j=1}^{m} g_{ij} f_{ji} \\ &= \sum_{i=1}^{m} \sum_{j=1}^{m} f_{ji} g_{ij} = \sum_{i=1}^{m} \sum_{j=1}^{m} f_{ij} g_{ji} \end{aligned}$$

So we get $tr(MN) = tr(NM)$.

(ii) By making use of (i), we have that

$$\begin{aligned} tr(N^{-1}MN) &= tr\{(N^{-1}M)N\} = tr\{N(N^{-1}M)\} \\ &= tr\{(N^{-1}N)M\} = tr(I_m M) = trM \end{aligned}$$

as desired.

Exercises of § 2-2 (page 26)

[1]

$$\text{(i) } MM' = \begin{pmatrix} 2 & 1 & 3 \\ -1 & 1 & 2 \end{pmatrix} \begin{pmatrix} -1 & 2 \\ 0 & 3 \\ 1 & -1 \end{pmatrix} = \begin{pmatrix} 1 & 4 \\ 3 & -1 \end{pmatrix}$$

$$\therefore (MM')M'' = \begin{pmatrix} 1 & 4 \\ 3 & -1 \end{pmatrix} \begin{pmatrix} 2 \\ 5 \end{pmatrix} = \begin{pmatrix} 22 \\ 1 \end{pmatrix}.$$

By proposition 2.2.5, $M(M'M'') = (MM')M''$.

$$\text{(ii) } MM' = \begin{pmatrix} 3 & 1 \\ 4 & 2 \end{pmatrix} \begin{pmatrix} 1 & -1 \\ 0 & 1 \end{pmatrix} = \begin{pmatrix} 3 & -2 \\ 4 & -2 \end{pmatrix}$$

$$\therefore (MM')M'' = \begin{pmatrix} 3 & -2 \\ 4 & -2 \end{pmatrix} \begin{pmatrix} -2 & 4 \\ 3 & 5 \end{pmatrix} = \begin{pmatrix} -12 & 2 \\ -14 & 6 \end{pmatrix}.$$

By proposition 2.2.5, $M(M'M'') = (MM')M''$.

[2] $$(M + M')^2 = (M + M')(M + M') = \begin{pmatrix} 6 & -2 & 8 \\ 2 & -7 & 5 \\ 15 & -7 & 8 \end{pmatrix},$$

$$(M - M')^2 = (M - M')(M - M') = \begin{pmatrix} 6 & 10 & 2 \\ -6 & -5 & 1 \\ -3 & -15 & 8 \end{pmatrix}.$$

$$\because M + M' = \begin{pmatrix} 2 & -2 & 2 \\ 3 & 0 & -1 \\ 4 & 1 & 1 \end{pmatrix}, M - M' \begin{pmatrix} 2 & 4 & -2 \\ -1 & 0 & 1 \\ -4 & -1 & -1 \end{pmatrix}.$$

$$MM' = \begin{pmatrix} 2 & -6 & 3 \\ 0 & -3 & 2 \\ 0 & 0 & 0 \end{pmatrix}, M'M = \begin{pmatrix} -3 & 0 & 0 \\ 4 & 2 & 0 \\ 9 & 4 & 0 \end{pmatrix}.$$

[3] By direct calculation, we have $E_{ij} \cdot E_{hk} = E_{ik}$ if $j = h$, but otherwise $E_{ij} \cdot E_{hk} = \mathbb{O}$.

[4]

$$_t(MM') = {}_t\begin{pmatrix} M_1(M')^1 & \cdots & M_1(M')^\ell \\ \vdots & & \vdots \\ M_m(M')^1 & \cdots & M_m(M')^\ell \end{pmatrix}$$

if $M = (f_{ij})_{m\times n}$ and $M' = (f'_{jk})_{n\times \ell}$. Hence

$$_t(MM') = \begin{pmatrix} M_1(M')^1 & \cdots & M_m(M')^1 \\ \vdots & & \vdots \\ M_1(M')^\ell & \cdots & M_m(M')^\ell \end{pmatrix}.$$

[5] $(M_1 + M_2)(M_1 - M_2) = M_1^2 - M_2^2$ if and only if $M_1^2 - M_1M_2 + M_2M_1 - M_2^2 = M_1^2 - M_2^2$ if and only if $M_2M_1 = M_1M_2$.

[6] By proposition 2.2.5, we have

$$\begin{aligned} &_t(M_1M_2M_3) = {}_t\{(M_1M_2)M_3\} = {}_t\{M_1(M_2M_3)\} \\ &= {}_t(M_2M_3){}_tM_1 = ({}_tM_3\ {}_tM_2){}_tM_1 = {}_tM_3\ {}_tM_2\ {}_tM_1. \end{aligned}$$

[7]

$$M_1^2 = \begin{pmatrix} 1 & 1 & 1 \\ 0 & 1 & 1 \\ 0 & 0 & 1 \end{pmatrix}^2 = \begin{pmatrix} 1 & 2 & 3 \\ 0 & 1 & 2 \\ 0 & 0 & 1 \end{pmatrix},$$

$$M_1^3 = \begin{pmatrix} 1 & 1 & 1 \\ 0 & 1 & 1 \\ 0 & 0 & 1 \end{pmatrix} \begin{pmatrix} 1 & 2 & 3 \\ 0 & 1 & 2 \\ 0 & 0 & 1 \end{pmatrix} = \begin{pmatrix} 1 & 3 & 6 \\ 0 & 1 & 3 \\ 0 & 0 & 1 \end{pmatrix};$$

$$M_2^2 = \begin{pmatrix} 0 & 1 & 1 \\ 0 & 0 & 1 \\ 0 & 0 & 0 \end{pmatrix}^2 = \begin{pmatrix} 0 & 0 & 1 \\ 0 & 0 & 0 \\ 0 & 0 & 0 \end{pmatrix},$$

$$M_2^3 = \begin{pmatrix} 0 & 1 & 1 \\ 0 & 0 & 1 \\ 0 & 0 & 0 \end{pmatrix} \begin{pmatrix} 0 & 0 & 1 \\ 0 & 0 & 0 \\ 0 & 0 & 0 \end{pmatrix} = \begin{pmatrix} 0 & 0 & 0 \\ 0 & 0 & 0 \\ 0 & 0 & 0 \end{pmatrix}.$$

[8] $\forall v = (f_1, \cdots, f_\ell), v' = (f'_1, \cdots, f'_\ell) \in F^\ell$, we may write

$$\begin{aligned} v \cdot v' &= \sum_{i=1}^{\ell} f_i f'_i = (f_1, \cdots, f_\ell)_{1\times\ell} \cdot \begin{pmatrix} f'_1 \\ \vdots \\ f'_\ell \end{pmatrix}_{\ell\times 1} \\ &= v_t(v') \in F. \end{aligned}$$

Exercises of § 2-3 (page 31)

[1]

$$\text{column rank} \begin{pmatrix} f_{11} & & \bigcirc \\ \vdots & \ddots & \\ f_{m1} & \cdots & f_{mm} \end{pmatrix}$$

$$= \text{row rank} \begin{pmatrix} f_{11} & & \bigcirc \\ \vdots & \ddots & \\ f_{m1} & \cdots & f_{mm} \end{pmatrix},$$

$$\text{column rank} \begin{pmatrix} f_{11} & \cdots & f_{1m} \\ & \ddots & \vdots \\ \bigcirc & & f_{mm} \end{pmatrix}$$

$$= \text{row rank} \begin{pmatrix} f_{11} & \cdots & f_{1m} \\ & \ddots & \vdots \\ \bigcirc & & f_{mm} \end{pmatrix}$$

by mathematical induction on $diag(f_{11}, \cdots, f_{mm})$.

[2]

(i) $\text{column rank} \begin{pmatrix} 7 & 2 & -1 \\ 4 & 1 & 2 \end{pmatrix} = 2 = \text{row rank} \begin{pmatrix} 7 & 2 & -1 \\ 4 & 1 & 2 \end{pmatrix}$

(ii) $\text{column rank} \begin{pmatrix} 7 & 3 \\ 3 & 4 \\ -4 & 3 \end{pmatrix} = 2 = \text{row rank} \begin{pmatrix} 7 & 3 \\ 3 & 4 \\ -4 & 3 \end{pmatrix}$

(iii) $\text{column rank} \begin{pmatrix} -7 & 8 & 3 \\ 1 & 2 & 5 \\ 0 & 0 & 4 \end{pmatrix} = 3 = \text{row rank} \begin{pmatrix} -7 & 8 & 3 \\ 1 & 2 & 5 \\ 0 & 0 & 4 \end{pmatrix}$

[3]

$$\text{(i)} \begin{cases} 7x_1 + 2x_2 - x_3 = 0 & \cdots (1) \\ 4x_1 + x_2 + 2x_3 = 0 & \cdots (2) \end{cases}$$

From (1), $x_3 = 7x_1 + 2x_2$. Substituting this for x_3 in (2), we have $4x_1 + x_2 + 2(7x_1 + 2x_2) = 18x_1 + 5x_2 = 0$, so that $x_2 = -\frac{18}{5}x_1$ and $x_3 = 7x_1 + 2(-\frac{18}{5}x_1) = -\frac{1}{5}x_1$. Hence $x_1 : x_2 : x_3 = 1 : (-\frac{18}{5}) : (-\frac{1}{5}) = -5 : 18 : 1$ if $\text{char}F = p \neq 5$.

It follows that $\{f(-5, 18, 1) \mid \forall f \in F\}$ is the solution space of the given system. Evidently the solution space has 1 as its dimension.

$$\text{(ii)} \begin{cases} 7x_1 + 3x_2 = 0 & \cdots (1) \\ 3x_1 + 4x_2 = 0 & \cdots (2) \\ -4x_1 + 3x_2 = 0 & \cdots (3) \end{cases}$$

From (1), $x_2 = -\frac{7}{3}x_1$. Substituting this for x_2 in (2) and (3), we have $3x_1 + 4(-\frac{7}{3}x_1) = -\frac{19}{3}x_1 = 0$ and $-4x_1 + 3(-\frac{7}{3}x_1) = -\frac{33}{3}x_1 = -11x_1 = 0$ if char$F = p \neq 3$. So $x_1 = 0 = x_2$. In this case, the solution space has 0 as its dimension.

$$\text{(iii)}\begin{cases} -7x_1 + 8x_2 + 3x_3 = 0 & \cdots (1) \\ x_1 + 2x_2 + 5x_3 = 0 & \cdots (2) \\ 4x_3 = 0 & \cdots (3) \end{cases}$$

Let $p =$char$F \neq 2, 11$. Similarly to (i) and (ii) above, we have $x_3 = x_2 = x_1 = 0$. So the dimension of the solution space is zero.

[4]

$$\text{(i)}\begin{cases} 7x_1 + 2x_2 - x_3 = 2 & \cdots (1) \\ 4x_1 + x_2 + 2x_3 = 1 & \cdots (2) \end{cases}$$

Any solution of the system of homogeneous linear equations is of the form $\{f(-5, 18, 1) \mid \forall f \in F\}$ by virtue of [3](i). Since $(x_1 = 0, x_2 = 1, x_3 = 0)$ is a particular solution of the given system of inhomogeneous linear equations, any solution of the given system is of the form $(0, 1, 0) + f(-5, 18, 1), \forall f \in F$ according to proposition 2.3.5.

$$\text{(ii)}\begin{cases} 7x_1 + 3x_2 = 2 & \cdots (1) \\ 3x_1 + 4x_2 = 1 & \cdots (2) \\ -4x_1 + 3x_2 = 0 & \cdots (3) \end{cases}$$

We must find a particular solution in this system, but substituting $x_1 = \frac{3}{4}x_2$ for x_1 (1),(2) yields an inconsistent system $\begin{cases} \frac{25}{4}x_2 = 1 \\ \frac{33}{4}x_2 = 2 \end{cases}$. So the solution set of the given system is empty.

$$\text{(iii)}\begin{cases} -7x_1 + 8x_2 + 3x_3 = 2 & \cdots(1) \\ x_1 + 2x_2 + 5x_3 = 0 & \cdots(2) \\ 4x_3 = -1 & \cdots(3) \end{cases}$$

If char$F = p \neq 2$, then $x_3 = -\frac{1}{4}$ by (3). Substituting $x_3 = -\frac{1}{4}$ for x_3 in (1),(2), we have

$$\begin{cases} -7x_1 + 8x_2 - \frac{3}{4} = 2 \\ 4x_1 + 8x_2 - 5 = 0. \end{cases}$$

So if $p \neq 11$, then $x_1 = \frac{9}{44}$ and $x_2 = \frac{46}{88}$. Hence $\exists!$ solution $(\frac{3^2}{2^2 \times 11}, -\frac{46}{2^3 \times 11}, -\frac{1}{4})$ for the given system.

[5]

(i)$S = \{(x_1, \cdots, x_n) \in F^n \mid (f_{ij})_{m\times n}\ {}_t(x_1, \cdots, x_n) = 0\}$.

(ii)$S = \{(x_1, \cdots, x_n) \in F^n \mid (f_{11}, \cdots, f_{1n}) \perp (x_1, \cdots, x_n), \cdots,$
$(f_{m1}, \cdots, f_{mn}) \perp (x_1, \cdots, x_n)\}$.

(iii)$S = \{(x_1, \cdots, x_n) \in F^n \mid x_1\ {}_t(f_{11}, \cdots, f_{m1}) + \cdots +$
$x_n\ {}_t(f_{1n}, \quad \cdots, f_{mn}) = 0\}$.

Here (i),(ii), and (iii) are all equivalent to the solution set of the following system

$$\left\{\begin{array}{l} f_{11}x_1 + \cdots + f_{1n}x_n = 0 \\ f_{21}x_1 + \cdots + f_{2n}x_n = 0 \\ \vdots \qquad \vdots \quad \vdots \qquad \vdots \\ f_{m1}x_1 + \cdots + f_{mn}x_n = 0 \end{array}\right\}.$$

SKETCH OF ANSWERS TO EXERCISES

Exercises of § 3-1 (page 44)

[1] We multiply each row up to $(\ell - 1)$th row by x_1 and subtract it from the next row below starting from the bottom row, so that we have

$$V_\ell = \begin{vmatrix} 1 & 1 & \cdots & 1 \\ 0 & x_2 - x_1 & \cdots & x_\ell - x_1 \\ \vdots & \vdots & & \vdots \\ 0 & -x_1 x_2^{\ell-2} + x_2^{\ell-1} & \cdots & -x_1 x_\ell^{\ell-2} + x_\ell^{\ell-1} \end{vmatrix}$$

$$= \begin{vmatrix} (x_2 - x_1) & \cdots & (x_\ell - x_1) \\ (x_2 - x_1)x_2 & \cdots & (x_\ell - x_1)x_\ell \\ \vdots & & \vdots \\ (x_2 - x_1)x_2^{\ell-2} & \cdots & (x_\ell - x_1)x_\ell^{\ell-2} \end{vmatrix}$$

$$= (x_\ell - x_1)\cdots(x_2 - x_1) \begin{vmatrix} 1 & \cdots & 1 \\ x_2 & \cdots & x_\ell \\ \vdots & & \vdots \\ x_2^{\ell-2} & \cdots & x_\ell^{\ell-2} \end{vmatrix}$$

$$= (x_\ell - x_1)\cdots(x_2 - x_1)V_{\ell-1}.$$

Typeset by $\mathcal{AMS}$-TEX

By induction hypothesis, it follows that $V_\ell = (x_\ell - x_1) \cdots (x_2 - x_1) \prod_{j>i>1}(x_j - x_i) = \prod_{j>i}(x_j - x_i)$.

[2]

$$|xI_\ell| = D(xI_\ell) = D(diag(x, \cdots, x)) = x^\ell,$$

$$D(diag(x_{11}, x_{22}, \cdots, x_{\ell\ell})) = x_{11}x_{22}\cdots x_{\ell\ell}.$$

[3]

$$D\begin{pmatrix} f_{11} & f_{12} & \cdots & f_{1\ell} \\ 0 & f_{22} & \cdots & f_{2\ell} \\ \vdots & \vdots & \ddots & \vdots \\ 0 & 0 & \cdots & f_{\ell\ell} \end{pmatrix} = f_{11}f_{22}\cdots f_{\ell\ell}.$$

$$D\begin{pmatrix} f_{11} & 0 & \cdots & 0 \\ f_{21} & f_{22} & \cdots & 0 \\ \vdots & \vdots & \ddots & \vdots \\ \vdots & \vdots & & 0 \\ f_{m1} & f_{m2} & \cdots & f_{mm} \end{pmatrix} = f_{11}f_{22}\cdots f_{mm}.$$

$$D\begin{pmatrix} cos\theta & sin\theta \\ -\sin\theta & cos\theta \end{pmatrix} = cos^2\theta + sin^2\theta = 1.$$

[4] Add the first and the second columns to the fourth column and get

$$\begin{vmatrix} 1 & x & y & z+w \\ 1 & y & z & w+x \\ 1 & z & w & x+y \\ 1 & w & x & y+z \end{vmatrix} = \begin{vmatrix} 1 & x & y & x+y+z+w \\ 1 & y & z & y+z+w+x \\ 1 & z & w & z+w+x+y \\ 1 & w & x & w+x+y+z \end{vmatrix}$$

$$= (x+y+w+z)\begin{vmatrix} 1 & x & y & 1 \\ 1 & y & z & 1 \\ 1 & z & w & 1 \\ 1 & w & x & 1 \end{vmatrix} = 0.$$

[5]

$$\begin{aligned}
\frac{d}{dx}D(M) =& \frac{d}{dx}\Big(\sum_{\bar\pi} \varepsilon(\bar\pi) f_{\bar\pi(1),1}(x) f_{\bar\pi(2),2}(x) f_{\bar\pi(3),3}(x)\Big) \\
=& \sum_{\bar\pi} \varepsilon(\bar\pi) f'_{\bar\pi(1),1}(x) f_{\bar\pi(2),2}(x) f_{\bar\pi(3),3}(x) \\
&+ \sum_{\bar\pi} \varepsilon(\bar\pi) f_{\bar\pi(1),1}(x) f'_{\bar\pi(2),2}(x) f_{\bar\pi(3),3}(x) \\
&+ \sum_{\bar\pi} \varepsilon(\bar\pi) f_{\bar\pi(1),1}(x) f_{\bar\pi(2),2}(x) f'_{\bar\pi(3),3}(x) \\
=& \begin{vmatrix} f'_{11}(x) & f_{12}(x) & f_{13}(x) \\ f'_{21}(x) & f_{22}(x) & f_{23}(x) \\ f'_{31}(x) & f_{32}(x) & f_{33}(x) \end{vmatrix} \\
&+ \begin{vmatrix} f_{11}(x) & f'_{12}(x) & f_{13}(x) \\ f_{21}(x) & f'_{22}(x) & f_{23}(x) \\ f_{31}(x) & f'_{32}(x) & f_{33}(x) \end{vmatrix} \\
&+ \begin{vmatrix} f_{11}(x) & f_{12}(x) & f'_{13}(x) \\ f_{21}(x) & f_{22}(x) & f'_{23}(x) \\ f_{31}(x) & f_{32}(x) & f'_{33}(x) \end{vmatrix}.
\end{aligned}$$

[6] (i)

$$\begin{aligned}
\varepsilon(2\ 6\ 7\ 3\ 1\ 4\ 5) &= \varepsilon\{((6,7)(5,6)(4,7)(3,7)(1,6)(2,6))\}^{-1}; \\
\varepsilon(2\ 3\ 5\ 4\ 1) &= \varepsilon\{((3,5)(2,3)(1,2))\}^{-1} = -1; \\
\varepsilon(5\ 3\ 4\ 1\ 2\ 6) &= \varepsilon\{((5,4)(3,4)(2,3)(5,1))\}^{-1} = 1.
\end{aligned}$$

(ii)

$$\begin{vmatrix} 2 & 1 & 3 \\ -1 & 2 & 1 \\ 4 & 3 & 5 \end{vmatrix} = \begin{vmatrix} 2+(-1)\times 2 & 1+2\times 2 & 3+1\times 2 \\ -1 & 2 & 1 \\ 4 & 3 & 5 \end{vmatrix}$$

$$= \begin{vmatrix} 0 & 5 & 5 \\ -1 & 2 & 1 \\ 4 & 3 & 5 \end{vmatrix} = -5\begin{vmatrix} -1 & 1 \\ 4 & 5 \end{vmatrix} + 5\begin{vmatrix} -1 & 2 \\ 4 & 3 \end{vmatrix}$$

$$= -5(-9) + 5(-11) = -10;$$

$$\begin{vmatrix} -1 & 0 & 3 & 0 \\ 1 & 3 & 1 & 4 \\ 2 & 2 & 5 & 1 \\ 0 & 1 & 7 & 2 \end{vmatrix} = \begin{vmatrix} -1 & 0 & 3+(-1)\times 3 & 0 \\ 1 & 3 & 1+1\times 3 & 4 \\ 2 & 2 & 5+2\times 3 & 1 \\ 0 & 1 & 7+0\times 3 & 2 \end{vmatrix}$$

$$= \begin{vmatrix} -1 & 0 & 0 & 0 \\ 1 & 3 & 4 & 4 \\ 2 & 2 & 11 & 1 \\ 0 & 1 & 7 & 2 \end{vmatrix} = -\begin{vmatrix} 3 & 4 & 4 \\ 2 & 11 & 1 \\ 1 & 7 & 2 \end{vmatrix}$$

$$= -\begin{vmatrix} 3 & 4-4\times 1 & 4 \\ 2 & 11-1\times 1 & 1 \\ 1 & 7-2\times 1 & 2 \end{vmatrix} = -\begin{vmatrix} 3 & 0 & 4 \\ 2 & 10 & 1 \\ 1 & 5 & 2 \end{vmatrix}$$

$$= -3\begin{vmatrix} 10 & 1 \\ 5 & 2 \end{vmatrix} - 4\begin{vmatrix} 2 & 10 \\ 1 & 5 \end{vmatrix} = -3(15) - 4\times 0 = -45;$$

$$\begin{vmatrix} 3 & 5 & 7 \\ 1 & 5 & 8 \\ 1 & 2 & 8 \end{vmatrix} = \begin{vmatrix} 3 & 5 & 7 \\ 1-1\times 1 & 5-2\times 1 & 8-8\times 1 \\ 1 & 2 & 8 \end{vmatrix}$$

$$= \begin{vmatrix} 3 & 5 & 7 \\ 0 & 3 & 0 \\ 1 & 2 & 8 \end{vmatrix} = 3\begin{vmatrix} 3 & 7 \\ 1 & 8 \end{vmatrix} = 3(24-7) = 51.$$

[7] We shall prove the latter part of proposition 3.1.9 (iii) in detail. Let $M = (f_{ij})_{\ell\times\ell}$ with $f_{ij} \in F$ and let $M^i = M^j$. We then have

$$D(M) = \sum \varepsilon(k_1, k_2, \cdots, k_\ell) f_{k_1,1} f_{k_2,2} \cdots f_{k_i,i} \cdots f_{k_j,j} \cdots f_{k_\ell,\ell}.$$

For any term in this expression, say

$$\varepsilon(k_1, \cdots, k_i, \cdots, k_j, \cdots, k_\ell) f_{k_1,1} f_{k_2,2} \cdots f_{k_i,i} \cdots f_{k_j,j} \cdots f_{k_\ell,\ell}$$

there always appears the term

$$\varepsilon(k_1, \cdots k_{i-1}, k_j, k_{i+1}, \cdots, k_{j-1}, k_i, k_{j+1}, \cdots, k_\ell)$$
$$f_{k_1,1} \cdots f_{k_{i-1},i-1} f_{k_j,i} f_{k_{i+1},i+1} \cdots f_{k_{j-1},j-1} f_{k_i,j} f_{k_{j+1},j+1} \cdots f_{k_\ell,\ell}$$

with $f_{k_j,i} = f_{k_j,j}$ and $f_{k_i,j} = f_{k_i,i}$, which is just equal to

$$-\varepsilon(k_1, \cdots, k_i, \cdots, k_j, \cdots, k_\ell) f_{k_1,1} f_{k_2,2} \cdots f_{k_i,i} \cdots f_{k_j,j} \cdots f_{k_\ell,\ell}.$$

Hence $D(M) = 0$.

Exercises of § 3-2 (page 49)

[1] (i) By virtue of Cramer's formula, we have

$$x = \frac{\begin{vmatrix} 1 & -2 & 1 \\ 4 & 1 & -1 \\ 2 & -3 & 1 \end{vmatrix}}{\begin{vmatrix} -3 & -2 & 1 \\ -1 & 1 & -1 \\ -2 & -3 & 1 \end{vmatrix}}, y = \frac{\begin{vmatrix} -3 & 1 & 1 \\ -1 & 4 & -1 \\ -2 & 2 & 1 \end{vmatrix}}{\begin{vmatrix} -3 & -2 & 1 \\ -1 & 1 & -1 \\ -2 & -3 & 1 \end{vmatrix}}, z = \frac{\begin{vmatrix} -3 & -2 & 1 \\ -1 & 1 & 4 \\ -2 & -3 & 2 \end{vmatrix}}{\begin{vmatrix} -3 & -2 & 1 \\ -1 & 1 & -1 \\ -2 & -3 & 1 \end{vmatrix}}$$

Note that $\begin{vmatrix} -3 & -2 & 1 \\ -1 & 1 & -1 \\ -2 & -3 & 1 \end{vmatrix} = 5 \neq 0.$

(ii) We first compute the determinant of the coefficients of the given system:

$$\begin{vmatrix} 1 & \alpha & \alpha^2 \\ 1 & \beta & \beta^2 \\ 1 & \gamma & \gamma^2 \end{vmatrix} = \begin{vmatrix} 1 & 1 & 1 \\ \alpha & \beta & \gamma \\ \alpha^2 & \beta^2 & \gamma^2 \end{vmatrix} = (\gamma-\beta)(\beta-\alpha)(\gamma-\alpha) \neq 0$$

by Vandermonde in [1] of exercises §3 − 1. So

$$x_1 = \frac{\begin{vmatrix} \alpha^3 & \alpha & \alpha^2 \\ \beta^3 & \beta & \beta^2 \\ \gamma^3 & \gamma & \gamma^2 \end{vmatrix}}{\begin{vmatrix} 1 & \alpha & \alpha^2 \\ 1 & \beta & \beta^2 \\ 1 & \gamma & \gamma^2 \end{vmatrix}}, x_2 = \frac{\begin{vmatrix} 1 & \alpha^3 & \alpha^2 \\ 1 & \beta^3 & \beta^2 \\ 1 & \gamma^3 & \gamma^2 \end{vmatrix}}{\begin{vmatrix} 1 & \alpha & \alpha^2 \\ 1 & \beta & \beta^2 \\ 1 & \gamma & \gamma^2 \end{vmatrix}}, x_3 = \frac{\begin{vmatrix} 1 & \alpha & \alpha^3 \\ 1 & \beta & \beta^3 \\ 1 & \gamma & \gamma^3 \end{vmatrix}}{\begin{vmatrix} 1 & \alpha & \alpha^2 \\ 1 & \beta & \beta^2 \\ 1 & \gamma & \gamma^2 \end{vmatrix}}.$$

[2] (i) We compute the determinant of the coefficients of the given system:

$$\begin{vmatrix} 1 & 2 & -3 & 4 \\ -2 & 1 & -4 & -1 \\ 1 & 1 & 1 & 1 \\ -1 & -1 & -1 & 1 \end{vmatrix} = \begin{vmatrix} 1 & 2 & -3 & 4 \\ -2 & 1 & -4 & -1 \\ 1 & 1 & 1 & 1 \\ -1+1 & -1+1 & -1+1 & 1+1 \end{vmatrix}$$

$$= \begin{vmatrix} 1 & 2 & -3 & 4 \\ -2 & 1 & -4 & -1 \\ 1 & 1 & 1 & 1 \\ 0 & 0 & 0 & 2 \end{vmatrix} = 2\begin{vmatrix} 1 & 2 & -3 \\ -2 & 1 & -4 \\ 1 & 1 & 1 \end{vmatrix}$$

$$=2\begin{vmatrix} 1-1 & 2-1 & -3-1 \\ -2 & 1 & -4 \\ 1 & 1 & 1 \end{vmatrix} = 2\begin{vmatrix} 0 & 1 & -4 \\ -2 & 1 & -4 \\ 1 & 1 & 1 \end{vmatrix}$$

$$=2\{2\begin{vmatrix} 1 & -4 \\ 1 & 1 \end{vmatrix} + \begin{vmatrix} 1 & -4 \\ 1 & -4 \end{vmatrix}\} = 2\{2(1+4)+0\} = 20 \neq 0.$$

Hence

$$w = \frac{\begin{vmatrix} 1 & 2 & -3 & 4 \\ 0 & 1 & -4 & -1 \\ 0 & 1 & 1 & 1 \\ 3 & -1 & -1 & 1 \end{vmatrix}}{\begin{vmatrix} 1 & 2 & -3 & 4 \\ -2 & 1 & -4 & -1 \\ 1 & 1 & 1 & 1 \\ -1 & -1 & -1 & 1 \end{vmatrix}}, x = \frac{\begin{vmatrix} 1 & 1 & -3 & 4 \\ -2 & 0 & -4 & -1 \\ 1 & 0 & 1 & 1 \\ -1 & 3 & -1 & 1 \end{vmatrix}}{\begin{vmatrix} 1 & 2 & -3 & 4 \\ -2 & 1 & -4 & -1 \\ 1 & 1 & 1 & 1 \\ -1 & -1 & -1 & 1 \end{vmatrix}},$$

$$y = \frac{\begin{vmatrix} 1 & 2 & 1 & 4 \\ -2 & 1 & 0 & -1 \\ 1 & 1 & 0 & 1 \\ -1 & -1 & 3 & 1 \end{vmatrix}}{\begin{vmatrix} 1 & 2 & -3 & 4 \\ -2 & 1 & -4 & -1 \\ 1 & 1 & 1 & 1 \\ -1 & -1 & -1 & 1 \end{vmatrix}}, z = \frac{\begin{vmatrix} 1 & 2 & -3 & 1 \\ -2 & 1 & -4 & 0 \\ 1 & 1 & 1 & 0 \\ -1 & -1 & -1 & 3 \end{vmatrix}}{\begin{vmatrix} 1 & 2 & -3 & 4 \\ -2 & 1 & -4 & -1 \\ 1 & 1 & 1 & 1 \\ -1 & -1 & -1 & 1 \end{vmatrix}}.$$

(ii) Similar as in (i).

[3] Put $\begin{vmatrix} X & Y \\ O & Z \end{vmatrix} =: |(f_{ij})_{2\ell\times 2\ell}|$. By definition and conditions, we have

$$\begin{aligned} \begin{vmatrix} X & Y \\ O & Z \end{vmatrix} &= \sum_{(i_1,\cdots,i_{2\ell})} \varepsilon(i_1,\cdots,i_{2\ell}) f_{i_1,1}\cdots f_{i_\ell,\ell}\cdots f_{i_{2\ell},2\ell} \\ &= \sum_{\substack{1\le i_1,\cdots,i_\ell\le\ell \\ \ell+1\le i_{\ell+1},\cdots,i_{2\ell}\le 2\ell}} \varepsilon(i_1,\cdots,i_{2\ell}) f_{i_1,1}\cdots f_{i_\ell,\ell}\cdots f_{i_{2\ell},2\ell} \\ &= \sum_{\ell+1\le i_{\ell+1},\cdots,i_{2\ell}\le 2\ell} \varepsilon(i_{\ell+1},\cdots,i_{2\ell}) f_{i_{\ell+1},\ell+1}\cdots f_{i_{2\ell},2\ell} \\ &\quad \sum_{1\le i_1,\cdots,i_\ell\le\ell} \varepsilon(i_1,\cdots,i_\ell) f_{i_1,1}\cdots f_{i_\ell,\ell} \\ &= |Z|\cdot|X|. \end{aligned}$$

Exercises of § 3-3 (page 53)

[1] (i)

$$D(M) = \begin{vmatrix} 1 & 1 & 2 \\ 1 & -3 & 0 \\ -2 & 1 & 4 \end{vmatrix} = -\begin{vmatrix} 1 & 2 \\ 1 & 4 \end{vmatrix} - 3\begin{vmatrix} 1 & 2 \\ -2 & 4 \end{vmatrix}$$
$$= -2 - 3(8) = -26 \neq 0$$

$$M_{11} = \begin{vmatrix} -3 & 0 \\ 1 & 4 \end{vmatrix} = -12,\ M_{12} = -\begin{vmatrix} 1 & 0 \\ -2 & 4 \end{vmatrix} = -4,$$
$$M_{13} = \begin{vmatrix} 1 & -3 \\ -2 & 1 \end{vmatrix} = -5,\quad M_{21} = -\begin{vmatrix} 1 & 2 \\ 1 & 4 \end{vmatrix} = -2,$$
$$M_{22} = \begin{vmatrix} 1 & 2 \\ -2 & 4 \end{vmatrix} = 8,\quad M_{23} = -\begin{vmatrix} 1 & 1 \\ -2 & 1 \end{vmatrix} = -3,$$
$$M_{31} = \begin{vmatrix} 1 & 2 \\ -3 & 0 \end{vmatrix} = 6,\quad M_{32} = -\begin{vmatrix} 1 & 2 \\ 1 & 0 \end{vmatrix} = 2,$$
$$M_{33} = \begin{vmatrix} 1 & 1 \\ 1 & -3 \end{vmatrix} = -4.$$

$$\therefore M^{-1} = \frac{adj\,M}{D(M)} = \frac{{}_t(M_{ij})}{D(M)} = \frac{\begin{pmatrix} -12 & -2 & 6 \\ -4 & 8 & 2 \\ -5 & -3 & -4 \end{pmatrix}}{-26}.$$

(ii),(iii) are similar to (i).

[2] (i)

$$M = \begin{pmatrix} -4 & 2 \\ 3 & -1 \end{pmatrix};\ D(M) = \begin{vmatrix} -4 & 2 \\ 3 & -1 \end{vmatrix} = 4 - 6 = -2,$$
$$M_{11} = -1, M_{12} = -3, M_{21} = -2, M_{22} = -4.$$

$$\therefore M^{-1} = \frac{adj\,M}{D(M)} = \frac{\begin{pmatrix} -1 & -2 \\ -3 & -4 \end{pmatrix}}{-2}.$$

(ii),(iii) are similar to (i).

[3] Since $D(M) \neq 0$, M is invertible by proposition 3.3.2, so that $M\mathbb{X} = G$ implies $M^{-1}M\mathbb{X} = M^{-1}G$, i.e., $\mathbb{X} = M^{-1}G$ is obtained.

[4] ${}_tM({}_tM)^{-1} = id$, while ${}_t(M^{-1}){}_tM = {}_t(MM^{-1}) = {}_tid = id$ (id :=identity matrix). So $({}_tM)^{-1} = {}_t(M^{-1})$ by proposition 2.2.7.

[5] (i)

$$\begin{Bmatrix} 2x + y = 0 \\ 4x + 3y = -1 \end{Bmatrix} \iff \begin{pmatrix} 2 & 1 \\ 4 & 3 \end{pmatrix}\begin{pmatrix} x \\ y \end{pmatrix} = \begin{pmatrix} 0 \\ -1 \end{pmatrix}.$$

$$D\begin{pmatrix} 2 & 1 \\ 4 & 3 \end{pmatrix} = 2 \neq 0, \text{ so } \begin{pmatrix} 2 & 1 \\ 4 & 3 \end{pmatrix}^{-1} \text{ exists, In fact}$$

$$\begin{pmatrix} 2 & 1 \\ 4 & 3 \end{pmatrix}^{-1} = \frac{\begin{pmatrix} 3 & -1 \\ -4 & 2 \end{pmatrix}}{2} = \begin{pmatrix} \frac{3}{2} & -\frac{1}{2} \\ -2 & 1 \end{pmatrix}.$$

Hence we have

$$\begin{pmatrix} 2 & 1 \\ 4 & 3 \end{pmatrix}\begin{pmatrix} x \\ y \end{pmatrix} = \begin{pmatrix} 0 \\ -1 \end{pmatrix} \iff \begin{pmatrix} 2 & 1 \\ 4 & 3 \end{pmatrix}^{-1}\begin{pmatrix} 2 & 1 \\ 4 & 3 \end{pmatrix}\begin{pmatrix} x \\ y \end{pmatrix}$$
$$= \begin{pmatrix} 2 & 1 \\ 4 & 3 \end{pmatrix}^{-1}\begin{pmatrix} 0 \\ -1 \end{pmatrix} \iff \begin{pmatrix} x \\ y \end{pmatrix} = \begin{pmatrix} \frac{3}{2} & -\frac{1}{2} \\ -2 & 1 \end{pmatrix}\begin{pmatrix} 0 \\ -1 \end{pmatrix}$$
$$= \begin{pmatrix} \frac{1}{2} \\ -1 \end{pmatrix},$$

$$\text{i.e., } \begin{Bmatrix} x = \dfrac{1}{2} \\ y = -1 \end{Bmatrix}. \text{ (ii) and (iii) are similar to (i).}$$

[6]

$$(MM')(M')^{-1}M^{-1} = MIM^{-1} = MM^{-1} = I,$$
$$(MM')(MM')^{-1} = I.$$

Exercises of § 3-4 (page 59)

[1] We shall do by induction on $m \times m$. First, 1×1 case is evident. Next we assume that the inequality given is right for $(m-1) \times (n-1)$ case. Now for the $m \times n$ case, we reduce $M + M'$ to the form $\begin{pmatrix} 1 & 0 & \cdots & 0 \\ 0 & g_{22} & \cdots & g_{2n} \\ \vdots & \vdots & & \vdots \\ 0 & g_{m2} & \cdots & g_{mn} \end{pmatrix} =$ $\widehat{M} + \widehat{M'}$ by using ER(i) through EC(iii), where $rank\widehat{M} = rankM$ and $rank\widehat{M'} = rankM'$.

Write $(g_{ij})_{(m-1)\times(n-1)} = \widehat{m}_{11} + \widehat{m}'_{11}$ for (g_{ij}) with $\widehat{m}_{11}$ ($\widehat{m}'_{11}$ respectively) the (1,1)-th minor matrix of $\widehat{M}$ ($\widehat{M'}$ respectively). By induction hypothesis we have $rank(g_{ij})$ $\leq rank\ \widehat{m}_{11} + rank\ \widehat{m}'_{11}$.

So $rank(M + M') = 1 + rank(g_{ij}) \leq 1 + rank\ \widehat{m}_{11} + rank\ \widehat{m}'_{11} \leq rank\widehat{M} + rank\widehat{M'}$ since we may assume (1,1)-th component(or entry) of $\widehat{M}$ is nonzero.

$\therefore\ rank(M + M') \leq rank\widehat{M} + rank\widehat{M'} \leq rankM + rankM'$.

[2]

$$
\text{(ii)} \begin{pmatrix} 3 & 5 & -2 \\ 2 & -1 & 3 \\ 7 & 2 & 1 \end{pmatrix} \xrightarrow{\text{EC (iii)}} \begin{pmatrix} 3+5\times 2 & 5 & -2 \\ 2+(-1)\times 2 & -1 & 3 \\ 7+2\times 2 & 2 & 1 \end{pmatrix}
$$

$$
= \begin{pmatrix} 13 & 5 & -2 \\ 0 & -1 & 3 \\ 11 & 2 & 1 \end{pmatrix} \xrightarrow{\text{EC (iii)}} \begin{pmatrix} 13 & 5 & -2+5\times 3 \\ 0 & -1 & 3+(-1)\times 3 \\ 11 & 2 & 1+2\times 3 \end{pmatrix}
$$

$$
= \begin{pmatrix} 13 & 5 & 13 \\ 0 & -1 & 0 \\ 11 & 2 & 7 \end{pmatrix}.
$$

$$
D \begin{pmatrix} 13 & 5 & 13 \\ 0 & -1 & 0 \\ 11 & 2 & 7 \end{pmatrix} = - \begin{vmatrix} 13 & 13 \\ 11 & 7 \end{vmatrix} \neq 0.
$$

$$
\therefore rank \begin{pmatrix} 3 & 5 & -2 \\ 2 & -1 & 3 \\ 7 & 2 & 1 \end{pmatrix} = 3
$$

by corollarly 3.4.6.

(iv) By proposition 3.4.3,

$$
rank \begin{pmatrix} 1 & -2 & 3 & -4 \\ 5 & 6 & 7 & -8 \end{pmatrix} \leq 2.
$$

By the way, $rank \begin{pmatrix} 1 & -2 \\ 5 & 6 \end{pmatrix} = 2$, since $\begin{vmatrix} 1 & -2 \\ 5 & 6 \end{vmatrix} = 16 \neq 0$.

$$
\therefore rank \begin{pmatrix} 1 & -2 & 3 & -4 \\ 5 & 6 & 7 & -8 \end{pmatrix} = 2.
$$

Exercises of § 4-1 (page 66)

[1] $\forall r_1, r_2 \in \mathbb{R}, \forall f_1, f_2 \in V, \frac{d}{dx}(r_1 f_1 + r_2 f_2) = \frac{d}{dx}(r_1 f_1) + \frac{d}{dx}(r_2 f_2) = r_1 \frac{df_1}{dx} + r_2 \frac{df_2}{dx}$.

$\therefore \frac{d}{dx} : V \to V$ is an operator.

[2] $\forall r_1, r_2 \in \mathbb{R}, \forall f_1, f_2 \in V', \frac{d^n}{dx^n}(r_1 f_1 + r_2 f_2) = \frac{d^n}{dx^n}(r_1 f_1) + \frac{d^n}{dx^n}(r_2 f_2) = r_1 \frac{d^n f_1}{dx^n} + r_2 \frac{d^n f_2}{dx^n}$.

Hence $\frac{d^n}{dx^n}$ also becomes linear as in [1].

[3] $\forall r_1, r_2 \in \mathbb{R}$, $\forall$ integrable functions f_1, f_2 on $\mathbb{R}$, we have $\int (r_1 f_1 + r_2 f_2) dx = \int r_1 f_1 dx + \int r_2 f_2 dx = r_1 \int f_1 dx + r_2 \int f_2 dx$. So $\int$ becomes linear.

[4] If L^{-1} exists as an inverse mapping of an operator L, then

$$
\begin{aligned}
(L^{-1})^m \circ L^m &= (\underbrace{L^{-1} \circ \cdots \circ L^{-1}}_{m-times}) \circ (\underbrace{L \circ \cdots \circ L}_{m-times}) \\
&= (\underbrace{L^{-1} \circ \cdots \circ L^{-1}}_{(m-1)-times}) \circ (L^{-1} \circ L) \circ (\underbrace{L \circ L \circ \cdots \circ L}_{(m-1)-times}) \\
&= (L^{-1} \circ \cdots \circ L^{-1}) \circ id \circ (L \circ \cdots \circ L) \\
&= (L^{-1} \circ \cdots \circ L^{-1}) \circ (L \circ \cdots \circ L) \\
&= \cdots = L^{-1} \circ id \circ L = L^{-1} \circ L = id.
\end{aligned}
$$

Similarly $L^m \circ (L^{-1})^m = id$. Hence L^m is invertible.

[5] ImL becomes a subspace of V. If $\{v_1, \cdots, v_\ell\}$ is a basis of V, then $Lv_1, \cdots, Lv_\ell$ are also independent in V. ($\because$

$f_1Lv_1 + \cdots + f_\ell Lv_\ell = 0 \Rightarrow L(f_1v_1) + \cdots + L(f_\ell v_\ell) = L(\sum_{i=1}^{\ell} f_iv_i) = 0$. Hence $\sum_i f_iv_i = 0$ since L is injective. But then $\forall f_i = 0$.) So $dimImL = \ell = dimV$. $\therefore ImL = V$. $\therefore L$ is bijective.

[6] Define a mapping $i : F \longrightarrow F^2$ by $i(f) = (f, 0)\ \forall f \in F$. We then easily see that i is injective as a linear map. But i is not onto since $i^{-1}(\{(0,1)\}) = \emptyset$.

Exercises of § 4-2 (page 72)

[1] $\forall v_1, v_2 \in L^{-1}(S'), L(sv_1 + (1-s)v_2) = L(sv_1) + L((1-s)v_2) = sL(v_1) + (1-s)L(v_2) \in S'$ since S' is a convex set in V', where $0 \leq s \leq 1$. $\therefore sv_1 + (1-s)v_2 \in L^{-1}(S')$.
$\therefore L^{-1}(S')$ is also convex in V'.

[2] $\forall v, w \in S_1 \cap S_2, sv + (1-s)w \in S_1$ and $sv + (1-s)w \in S_2$ since S_1 and S_2 are convex.
$\therefore sv + (1-s)w \in S_1 \cap S_2$. So $S_1 \cap S_2$ is convex.

[3] $\forall v_1, v_2 \in L^{-1}(\mathbb{Z}^{<0}), 0 \leq \forall s \leq 1$, we have $L(sv_1 + (1-s)v_2) = sL(v_1) + (1-s)L(v_2) < 0$ since $L(v_1) < 0$ and $L(v_2) < 0$. Hence $L^{-1}(\mathbb{Z}^{<0})$ is a convex set.

[4] Note that all $P_i's$ are in S. Since S is convex, the convex polyhedron determined by vertices $P_1, P_2, \cdots, P_\ell$ is included in S. Evidently the boundary of this polyhedron is equal to the boundary of S considering the definition of S and $\sum_{i=1}^{\ell} s_iP_i = (1-s_\ell)\sum_{i=1}^{\ell-1} \frac{s_i}{1-s_\ell}P_i + s_\ell P_\ell$ for $s_\ell \neq 1$.

Exercises of § 4-3 (page 79)

[1] From remark 4.3.5, we have $A^{\mathcal{B}}_{\mathcal{B}'}(id)X_{\mathcal{B}}(v) = X_{\mathcal{B}'}(v)$. If $\mathcal{B} = \mathcal{B}'$, then $A^{\mathcal{B}}_{\mathcal{B}}(id)X_{\mathcal{B}}(v) = X_{\mathcal{B}}(v)$.

$\therefore A^{\mathcal{B}}_{\mathcal{B}}(id) = I_m$ since $dim_F V = m$.

[2]

$$A^{\mathcal{B}_1}_{\mathcal{B}_2}(L) = \begin{pmatrix} 2 & -1 \\ 4 & 7 \\ -3 & 8 \end{pmatrix}.$$

[3]

$$\begin{aligned} L(U_1) &= cos\theta U_1 + sin\theta U_2, \\ L(U_2) &= cos(\frac{\pi}{2} + \theta)U_1 + sin(\frac{\pi}{2} + \theta)U_2 \\ &= -sin\theta U_1 + cos\theta U_2, \end{aligned}$$

Hence $A^{\mathcal{B}}_{\mathcal{B}}(L) = \begin{pmatrix} cos\theta & -sin\theta \\ sin\theta & cos\theta \end{pmatrix}$.

[4]

$$A^{\mathcal{B}}_{\mathcal{B}}(L) = \begin{pmatrix} f_{11} & \cdots & f_{n1} \\ \vdots & \cdots & \vdots \\ f_{1n} & \cdots & f_{nn} \end{pmatrix},$$

$$A^{\mathcal{B}'}_{\mathcal{B}}(id) = \begin{pmatrix} f_{11} & \cdots & f_{n1} \\ \vdots & \cdots & \vdots \\ f_{1n} & \cdots & f_{nn} \end{pmatrix}.$$

[5]

$$\forall(f_1,\cdots,f_m),(f_1',\cdots,f_m')\in F^m,$$
$$\pi_i\{(f_1,\cdots,f_m)+(f_1',\cdots,f_m')\}$$
$$=\pi_i(f_1+f_1',\cdots,f_m+f_m')$$
$$=f_i+f_i'=\pi_i(f_1,\cdots,f_m)+\pi_i(f_1',\cdots,f_m');$$

$$\forall f\in F,\forall(f_1,\cdots,f_m)\in F^m,\pi_i\{f(f_1,\cdots,f_m)\}$$
$$=\pi_i(ff_1,\cdots,ff_m)=ff_i=f\pi_i(f_1,\cdots,f_m).$$

$\therefore \forall\pi_i$ for $1\le i\le m$ becomes linear.

[6] $\forall X\in V,(L_1+L_2)X=L_1X+L_2X$ and $L_1+L_2\in LM(V,V')$. Similarly $\forall f\in F,\forall\in V,(fL_1)X=f(L_1X)$ and $fL_1\in LM(V,V')$. Since $LM(V,V')\cong M_{m\times n}(F)$ meaning that $LM(V,V')$ is isomorphic to $M_{m\times n}(F)$, it follows that $A^{\mathcal{B}}_{\mathcal{B}'}(L_1+L_2)=A^{\mathcal{B}}_{\mathcal{B}'}(L_1)+A^{\mathcal{B}}_{\mathcal{B}'}(L_2)$ and $A^{\mathcal{B}}_{\mathcal{B}'}(fL_1)=fA^{\mathcal{B}}_{\mathcal{B}'}(L_1)$.

SKETCH OF ANSWERS TO EXERCISES

Exercises of § 5-1 (page 91)

[1] (i) We let v' be an element of V satisfying $\|v'\| \neq 0$. $\forall v \in V$, we have a unique number $c \in \mathbb{C}$ satisfying v' is perpendicular to $v - cv'$. In fact, we may put $c = \dfrac{< v, v' >}{< v', v' >}$ using $< v - cv', v' >= 0$.

Here c is usually called the component of v along v' and cv' the projection of v along v'. Because $(v - cv') \perp cv'$, $\|v\|^2 = \|(v - cv') + cv'\|^2 =< (v - cv') + cv', (v - cv') + cv' >= \|v - cv'\|^2 + |c|^2\|v'\|^2$ holds.

In particular if we let $v' = u$(a unit vector), we then have $\|v\|^2 = \|v - cu\|^2 + |c|^2$, so that $|c|^2 \leq \|v\|^2$,i.e., $|c| \leq \|v\|$. Now for any nonzero vector v', we have a unit vector $\dfrac{v'}{\|v'\|}$. From what we have just seen, it follows that $\left| < v, \dfrac{v'}{\|v'\|} > \right| \leq \|v\|$, namely $| < v, v' > | \leq \|v'\|\|v\|$. In case $v' = 0$, our inequality $\leq$ is evident.

(ii) If $v = 0$, then $< v, v >=< v, v + v >=< v, v > + < v, v >$, so that $< v, v >= 0 = \|v\|^2$. Hence $\forall v \in$

Typeset by $\mathcal{A}\mathcal{M}\mathcal{S}$-TEX

$V, \|v\| \geq 0$. Because $v \neq 0 \Rightarrow \|v\| > 0$, we always have $v = 0$ if and only if $\|v\| = 0$.

(iii) $\|cv\|^2 =< cv, cv >$

$= c \cdot \bar{c} < v, v >= |c|^2 \|v\|^2, \|cv\| \geq 0.$

$\therefore \|cv\| = |c| \|v\|.$

(iv) Using (i), we have

$$\|v + v'\|^2 =< v + v', v + v' >$$

$$=< v, v > + < v', v' > + < v, v' > + < v'.v >$$

$$\leq \|v\|^2 + \|v'\|^2 + 2| < v, v' > |$$

$$\leq \|v\|^2 + \|v'\|^2 + 2\|v\|\|v'\| = (\|v\| + \|v'\|)^2.$$

$$\therefore \|v + v'\| \leq \|v\| + \|v'\|.$$

[2] The proof is exactly the same as that of theorem 5.1.13 except that $<, >$ should indicate a positive definite hermitian product.

[3] (i) If $S = \mathbb{O}$ or $S = V$, the equality is trivial. So we assume $S \neq \mathbb{O}$ and $S \neq V$. The rest of this proof is exactly the same as that of theorem 5.1.14 except that $<, >$ should denote a positive definite hermitian product this time.

(ii) Since $S \cap S^\perp = \{\mathbb{O}\}$ and since $dimS + dimS^\perp = dimV$ by (i), we have $V = S \oplus S^\perp$ by virtue of proposition 1.4.4.

[4] (i) $\|v+v'\|^2 = < v+v', v+v' > = < v,v > + < v',v' > + < v,v' > + < v',v > = \|v\|^2 + \|v'\|^2$ since v is perpendicular to v'.

(ii) $\|v+v'\|^2 + \|v-v'\|^2 = < v+v', v+v' > + < v-v', v-v' > = < v,v > + < v',v' > + < v,v > + < v',v' > = 2\|v\|^2 + 2\|v'\|^2$ because $v \perp v'$ holds.

[5] (i)

$$\begin{aligned} < f_2, f_1 > &= \int_{-\pi}^{\pi} f_2(x)\overline{f_1(x)}dx \\ &= \int_{-\pi}^{\pi} \overline{f_1(x)\overline{f_2(x)}}dx = \overline{\int_{-\pi}^{\pi} f_1(x)\overline{f_2(x)}dx} \\ &= \overline{< f_1, f_2 >}. \end{aligned}$$

(ii) $\forall f_1, f_2, f_3 \in V$,

$$\begin{aligned} < f_1, f_2 + f_3 > &= \int_{-\pi}^{\pi} f_1(x)\overline{(f_2+f_3)(x)}dx \\ &= \int_{-\pi}^{\pi} f_1(x)\{\overline{f_2(x)} + \overline{f_3(x)}\}dx \\ &= \int_{-\pi}^{\pi} f_1(x)\overline{f_2(x)}dx + \int_{-\pi}^{\pi} f_1(x)\overline{f_3(x)}dx \\ &= < f_1, f_2 > + < f_1, f_3 > . \end{aligned}$$

(iii) $\forall c \in \mathbb{C}, \forall f_1, f_2 \in V$,

$$\begin{aligned} < cf_1, f_2 > &= \int_{-\pi}^{\pi} cf_1(x)\overline{f_2(x)}dx \\ &= c\int_{-\pi}^{\pi} f_1(x)\overline{f_2(x)}dx \\ &= c < f_1, f_2 > . \end{aligned}$$

(iv) For $f \in V$,

$$0 =< f, f > \iff \int_{-\pi}^{\pi} f(x)\overline{f(x)}dx = 0$$
$$\iff \int_{-\pi}^{\pi} \|f(x)\|^2 dx = 0$$
$$\iff \|f(x)\| = 0 \iff f(x) = 0.$$

Next $\forall f(\neq 0) \in V$,

$$< f, f >= \int_{-\pi}^{\pi} f(x)\overline{f(x)}dx = \int_{-\pi}^{\pi} \|f(x)\|^2 dx > 0.$$

$\therefore <,>$ becomes a positive definite hermitian product.

[6] (i) It is easy to show that $<,>$ becomes a symmetric bilinear form, namely a scalar product.

(ii) $\forall f \in V, < f, f >= \int_{-1}^{1} f(x)^2 dx \geq 0$ since $f(x)^2 \geq 0$. Now $< f, f >= 0 = \int_{-1}^{1} f(x)^2 dx$ if and only if $f(x) = 0$ on [-1,1].

Hence $<,>$ is rendered a positive definite scalar product, which implies that $<,>$ is nondegenerate.

[7] Every positive definite scalar product (respectively hermitian product) becomes a nondegenerate scalar product (respectively hermitian product) automatically.

Exercises of § 5-2 (page 96)

[1] Very clear from the definition 5.1.16.

[2] $\forall s \in S, \forall s' \in S^{\perp}, < s, s' >=< s', s >= 0$. Hence $S \subset (S^{\perp})^{\perp}$. However $dimS + dimS^{\perp} = dimS^{\perp} + dim(S^{\perp})^{\perp} = dimV$ by proposition 5.2.8. So $dim(S^{\perp})^{\perp} = dimS$.

Hence $S = (S^{\perp})^{\perp}$.

[3] By proposition 5.2.4, $dimS^* = dimS$. Since the usual dot product is nondegenerate, $dimS^{\perp} = dim\mathbb{R}^m - dimS = m - dimS$ by proposition 5.2.8.

[4] $\forall (x_1, \cdots, x_m), (x'_1, \cdots, x'_m) \in F^m, \forall f \in F$,

$$\begin{aligned} &P_i\{(x_1, \cdots, x_m) + (x'_1, \cdots, x'_m)\} \\ &= P_i(x_1 + x'_1, \cdots, x_m + x'_m) \\ &= x_i + x'_i \\ &= P_i(x_1, \cdots, x_m) + P_i(x'_1, \cdots, x'_m); \\ &P_i\{f(x_1, \cdots, x_m)\} \\ &= P_i(fx_1, \cdots, fx_m) = fx_i = fP_i(x_1, \cdots, x_m). \end{aligned}$$

Hence $P_i \ \forall i$ becomes a linear form on F^m.

[5] Denote by L the map $V \longrightarrow V^{**}$ given by $v \mapsto x_v$, where $x_v(w^*) = w^*(v) \ \forall w^* \in V^*$.

(i) L is $1-1$: If $L(v) = L(v')$ for some $v, v' \in V$, then $\forall w^* \in V^*, L(v)(w^*) = L(v')(w^*)$ holds, so that $w^*(v) = w^*(v')$. Considering a dual basis of V, we must have $v = v'$.

(ii) $\forall v, v' \in V, \forall w^* \in V^*$, we have that $x_{(v+v')}(w^*) = w^*(v + v') = w^*(v) + w^*(v') = (x_v + x_{v'})(w^*)$. Hence $L(v + v') = L(v) + L(v')$.

(iii) $\forall f \in F, \forall v \in V, \forall w^* \in V^*$, we have $x_{fv}(w^*) = w^*(fv) = f(w^*(v)) = f\{x_v(w^*)\} = (fx_v)(w^*)$. Hence $L(fv) = f \cdot L(v)$.

(iv) L is onto since V is finite dimensional.

$\therefore$ L is an isomorphism.

Exercises of § 5-3 (page 100)

[1] (i) $rank \begin{pmatrix} 1 & -1 & 1 \\ 1 & 1 & 0 \\ 0 & 1 & -1 \end{pmatrix} = 3$. Hence the dimension of the solution space equals $3 - 3 = 0$.

(ii) $rank \begin{pmatrix} -2 & 1 & 1 \\ 7 & 4 & -\pi \end{pmatrix} = 2$. Hence the dimension of the solution space equals $3 - 2 = 1$.

[2] (i) $rank \begin{pmatrix} 1 & -1 & 1 \\ 4 & -4 & 4 \end{pmatrix} = 1$ since $4(1,-1,1) - (4,-4,4) \equiv \mathbb{O}$. Hence the dimension of the solution space $= 3 - 1 = 2$.

(ii) $rank \begin{pmatrix} 3 & 4 & -5 \end{pmatrix} = 1$, so that the dimension of the solution space $3 - 1 = 2$.

[3] Note first that the solution space of the equation $M\mathbb{X} = \mathbb{O}$ equals the kernel of L_M. By the way $dimKerL_M = dimF^n - dimImL_M$ and dim(the solution space)$=n - rankM$. $\therefore$ $dimImL_M = rankM$.

[4] $\nexists$ orthogonal basis of [1](i);$\{(1, \dfrac{2\pi - 7}{4+\pi}, \dfrac{15}{4+\pi})\}$ is an orthogonal basis of [1](ii); [2](i) is similar to [2](ii);

$\{(4,-3,0), (5,0,3)\}$ is a basis of the solution space of [2](ii). We put $v_1 = (4,-3,0), v_2 = (5,0,3)$. Next we get an orthogonal basis $\{s_1, s_2\}$ using Gram-Schmidt process: $s_1 := v_1, s_2 := v_2 - \dfrac{< v_2, s_1 >}{< s_1, s_1 >} s_1$. Starting from $(3, 4, -5)$, we may also use the method of Examples 5.3.3.

[5] First consider $L_{M_1} \circ L_{M_2}$ and choose a basis of ImL_{M_2}.

Next extend this basis to a basis of $V = {}_tF^n$. We then see that $ImL_{M_2} = ImL_M$, where $M = \begin{pmatrix} I_r & 0 \\ 0 & 0 \end{pmatrix}$ for some $r \in \mathbb{Z}^{\geq 0}$, by virtue of corollary 4.3.9. Hence $rankM_1M_2 = rankM_2M_1 = min(rankM_1, rankM_2)$.

So $dimImL_{M_1} \circ L_{M_2} = dimImL_{M_2} \circ L_{M_1}$ holds.

Exercises of § 6-1 (page 103)

[1]

$$\begin{aligned} q(X) &= (x, y) \begin{pmatrix} f_{11} & f_{12} \\ f_{21} & f_{22} \end{pmatrix} \begin{pmatrix} x \\ y \end{pmatrix} \\ &= f_{11}x^2 + 2f_{12}xy + f_{22}y^2 \\ &= 4x^2 + 3xy - 6y^2. \end{aligned}$$

$$\therefore\ A = \begin{pmatrix} 4 & \frac{3}{2} \\ \frac{3}{2} & -6 \end{pmatrix} \text{ is the associated matrix.}$$

[2]

$$\begin{aligned} q(X) &= (x, y, z) \begin{pmatrix} f_{11} & f_{12} & f_{13} \\ f_{21} & f_{22} & f_{23} \\ f_{31} & f_{32} & f_{33} \end{pmatrix} \begin{pmatrix} x \\ y \\ z \end{pmatrix} \\ &= 3x^2 + 4xy + 5yz + 6xz + 7y^2 - 8z^2. \end{aligned}$$

So $f_{11} = 3, f_{22} = 7, f_{33} = -8, f_{12} = f_{21} = 2, f_{13} = f_{31} = 3, f_{23} = f_{32} = \frac{5}{2}$.

[3] We let q be the quadratic form determined by $\beta_1 + \beta_2$. We then have $\forall v \in V, (\beta_1 + \beta_2)(v, v) = \beta_1(v, v) + \beta_2(v, v) = q_1(v) + q_2(v) = (q_1 + q_2)(v) = q(v)$.

$\therefore\ q = q_1 + q_2$.

[4] Since $q(X) = {}^tXAX = \sum_{i,j=1}^n f_{ij}x_ix_j = \sum_{i=1}^n x_i^2$, we have $\begin{cases} f_{ij} = 0 \text{ if } i \neq j; \\ f_{ij} = 1 \text{ if } i = j. \end{cases}$

Exercises of § 6-2 (page 106)

[1] Same proof as that of proposition 6.2.3 except that r_i may belong to any field $\mathbb{F}$.

[2] (i) $\{(1,0), (0,1)\}$: a basis of $\mathbb{F}^2$. Since $< (1,0), (1,0) > \neq 0, < (0,1), (0,1) > \neq 0$, we may orthogonalize this basis using Gram-Schmidt process. Put $s_1 := (1,0)$ at first. Next we have $s_2 := (0,1) - c_1s_1 = (0,1) - \frac{< (0,1), s_1 >}{< s_1, s_1 >}s_1$ $= (0,1) - \frac{3}{1}(1,0) = (0,1) - (3,0) = (-3,1)$. So $\{s_1, s_2\}$ is an orthogonal basis of $\mathbb{F}^2$.

Hence the index of nullity$= 0$ by proposition 6.2.3. since $< s_1, s_2 > \neq 0$ and $< s_2, s_2 > \neq 0$.

(ii) is similar to (i).

Exercises of § 6-3 (page 110)

[1] (i) Since $\begin{vmatrix} 1 & 1 \\ 1 & 0 \end{vmatrix} \neq 0$, $\{{}_t(1,1), {}_t(1,0)\}$ is a basis of V. Since $(1,1)\begin{pmatrix} -1 & 1 \\ 1 & -1 \end{pmatrix}\begin{pmatrix} 1 \\ 0 \end{pmatrix} = (0,0)\begin{pmatrix} 1 \\ 0 \end{pmatrix} = 0$, the above basis is an orthogonal basis of V.

(ii) $(1,1)\begin{pmatrix} -1 & 1 \\ 1 & -1 \end{pmatrix}\begin{pmatrix} 1 \\ 1 \end{pmatrix} = (0,0)\begin{pmatrix} 1 \\ 1 \end{pmatrix} = 0 \neq 1$, $(1,0)\begin{pmatrix} -1 & 1 \\ 1 & -1 \end{pmatrix}\begin{pmatrix} 1 \\ 0 \end{pmatrix} = (-1,1)\begin{pmatrix} 1 \\ 0 \end{pmatrix} = -1 \neq 1$. So $\{{}_t(1,1), {}_t(1,0)\}$ is not an orthonormal basis of V.

[2] (i) Put $v_1 := (1,0), v_2 := (0,1)$. By using Gram-Schmidt process, we have $s_1 := (1,0), s_2 := (0,1) - \frac{< v_2, s_1 >}{< s_1, s_1 >} s_1 = (-\frac{3}{2}, 1)$, which are orthogonal.

Hence the index of positivity $= 1$,

the index of negativity $= 1$,

the index of nullity $= 0$.

(ii) and (iii) are similar to (i).

[3] We consider the standard basis $\{U_1, \cdots, U_n\}$ consisting of standard unit vectors. Since $< U_i, U_i >= 1\ \forall i = 1, \cdots, n$, we have the index of positivity $=n$,

the index of negativity $= 0$,

the index of nullity $= 0$.

SKETCH OF ANSWERS TO EXERCISES

Exercises of § 7-1 (page 114)

[1] If we put $\mathcal{O} := \begin{pmatrix} 4 & 3 \\ 3 & -2 \end{pmatrix}$, then the quadratic form $q_{\mathcal{O}}$ determined by $\mathcal{O}$ is as follows. $\forall\ \mathbb{X} = {}_t(x_1, x_2)$,

$$\begin{aligned} q_{\mathcal{O}}(\mathbb{X}) &=< \mathcal{O}\mathbb{X}, \mathbb{X} >= {}_t(\mathcal{O}\mathbb{X}) I_2 \mathbb{X} \\ &= {}_t\mathbb{X}_t\mathcal{O}\mathbb{X} = (x_1, x_2) \begin{pmatrix} 4 & 3 \\ 3 & -2 \end{pmatrix} \begin{pmatrix} x_1 \\ x_2 \end{pmatrix} \\ &= (4x_1 + 3x_2, 3x_1 - 2x_2) \begin{pmatrix} x_1 \\ x_2 \end{pmatrix} \\ &= (4x_1 + 3x_2)x_1 + (3x_1 - 2x_2)x_2 \\ &= 4x_1^2 + 6x_1x_2 - 2x_2^2. \end{aligned}$$

[2] $\forall v, w \in V$,

$$\begin{aligned} < v, {}_t(\mathcal{O} + \mathcal{P})w > &=< (\mathcal{O} + \mathcal{P})v, w >=< \mathcal{O}v + \mathcal{P}v, w > \\ &=< \mathcal{O}v, w > + < \mathcal{P}v, w > \\ &=< v, {}_t\mathcal{O}w > + < v, {}_t\mathcal{P}w > \\ &=< v, ({}_t\mathcal{O} + {}_t\mathcal{P})w > . \end{aligned}$$

Typeset by $\mathcal{AMS}$-TEX

$\therefore\ {}_t(\mathcal{O}+\mathcal{P}) = {}_t\mathcal{O} + {}_t\mathcal{P}$;

$$\begin{aligned} < v, {}_t(f\mathcal{O})w > &=< (f\mathcal{O})v, w >=< f(\mathcal{O}v), w > \\ &= f < \mathcal{O}v, w >= f < v, {}_t\mathcal{O}w > \\ &=< v, f{}_t\mathcal{O}w > . \end{aligned}$$

So by nondegeneracy, ${}_t(f\mathcal{O}) = f{}_t\mathcal{O}$ is obtained as above;

$$\begin{aligned} < v, {}_t(\mathcal{O}\mathcal{P})w > &=< (\mathcal{O}\mathcal{P})v, w >=< \mathcal{O}(\mathcal{P}v), w > \\ &=< \mathcal{P}v, {}_t\mathcal{O}w >=< v, {}_t\mathcal{P}({}_t\mathcal{O}w) > \\ &=< v, ({}_t\mathcal{P}{}_t\mathcal{O})w > . \end{aligned}$$

Hence ${}_t(\mathcal{O}\mathcal{P}) = {}_t\mathcal{P}{}_t\mathcal{O}$.

[3] $-{}_t(A - {}_tA) = -{}_tA + A$. So $A - {}_tA$ is skew-symmetric. Let $A = a(A - {}_tA) + \mathbb{X}$ for some $a \in F$. We then have $X = (1-a)A + a{}_tA$, and so ${}_tX = (1-a){}_tA + aA$.

Putting $X = {}_tX$, we get $a = \dfrac{1}{2}$ and $X = \dfrac{1}{2}(A + {}_tA)$ which is symmetric. $\therefore\ A = \dfrac{1}{2}(A - {}_tA) + \dfrac{1}{2}(A + {}_tA)$.

[4] (i) Let $M_1 := (f_{ij})_{m\times m}, M_2 := (g_{ij})_{m\times m}$, $M_1 \cdot M_2 = (h_{ij})_{m\times m}$, and $M_2 \cdot M_1 = (h'_{ij})_{m\times m}$; we then have

$$tr(M_1 \cdot M_2) = \sum_{j=1}^{m} h_{jj} = \sum_{j=1}^{m}\sum_{i=1}^{m} f_{ji}g_{ij} \text{ and}$$

$$\begin{aligned} tr(M_2 \cdot M_1) &= \sum_{i=1}^{m} h_{ii} = \sum_{i=1}^{m}\sum_{j=1}^{m} g_{ij}f_{ji} \\ &= \sum_{j=1}^{m}\sum_{i=1}^{m} f_{ji}g_{ij}. \end{aligned}$$

Hence $tr(M_1M_2) = tr(M_2M_1)$.

(ii) Let $\mathcal{B}, \mathcal{B}'$ be two bases of $\mathbb{F}^m$; we then have

$$\begin{aligned} A^{\mathcal{B}'}_{\mathcal{B}'}(\mathcal{O}) &= A^{\mathcal{B}}_{\mathcal{B}'}(id) \circ A^{\mathcal{B}}_{\mathcal{B}}(\mathcal{O}) \circ A^{\mathcal{B}'}_{\mathcal{B}}(id) \\ &= \{A^{\mathcal{B}'}_{\mathcal{B}}(id)\}^{-1} \circ A^{\mathcal{B}}_{\mathcal{B}}(\mathcal{O}) \circ A^{\mathcal{B}'}_{\mathcal{B}}(id) \end{aligned}$$

by corollary 4.3.9. But then by(i), $trA^{\mathcal{B}'}_{\mathcal{B}'}(\mathcal{O}) = trA^{\mathcal{B}}_{\mathcal{B}}(\mathcal{O})$. Hence $tr\mathcal{O}$ is well-defined regardless of bases.

Since $trM = tr({}_tM)$, evidently $tr\mathcal{O} = tr_t\mathcal{O}$.

[5] By [4](i), we have

$$tr(A^{-1}MA) = tr\{A(A^{-1}M)\} = tr\{(AA^{-1})M\} = trM.$$

The latter part is also evident by [4](ii).

Exercises of § 7-2 (page 119)

[1] $\|v + w\|^2 =< v + w, v + w >=< v, v > + < v, w > + < w, v > + < w, w >=< v, v > + < w, w >= \|v\|^2 + \|w\|^2$.

[2] Let V be the given space. Let $< v, w >= 0$ for some $v \in V, \forall w \in V$; we then have $< v, v >= \|v\|^2 \geq 0$ and $= 0$ if and only if $v = 0$. Hence V is nondegenerate.

[3] Put $v_1 := (i, 2, -i, 1)$ and $v_2 := (1, i, 2, -i)$; then using

Gram-Schmidt process, we have

$$\begin{aligned} s_1 &:= v_1, \\ s_2 &:= v_2 - \frac{< v_2, s_1 >}{< s_1, s_1 >} s_1 \\ &= (1, i, 2, -i) - \frac{2i}{7}(i, 2, -i, 1) \\ &= \frac{1}{7}(9, 3i, 12, -9i). \end{aligned}$$

So the orthonormal basis is

$$\{\frac{s_1}{||s_1||}, \frac{s_2}{||s_2||}\}; ||s_1|| = \sqrt{7}, ||s_2|| = \frac{1}{7}\sqrt{9^2 - 9i^2 + 12^2 - 81i^2}.$$

[4] Proof of proposition 7.2.3 = Proof of [2], Exercises §5 − 1.

Proof of proposition 7.2.4 = Proof of [3], Exercises §5 − 1.

Proof of proposition 7.2.6 = Proof of proposition 7.1.1 except that $<, >$ is a hermitian form.

Proof of proposition 7.2.9 = Proof of proposition 7.1.3 except that $<, >$ is a hermitian form.

Exercises of § 7-3 (page 122)

[1] By proposition 7.3.3, we have ${}_t\mathcal{O}\mathcal{O} = id = \mathcal{O}_t\mathcal{O}$.

[2] If $\mathcal{O}, \mathcal{P}$ are unitary, then ${}_t(\mathcal{O}\mathcal{P})(\mathcal{O}\mathcal{P}) = {}_t\mathcal{P}_t\mathcal{O}(\mathcal{O}\mathcal{P}) = {}_t\mathcal{P}({}_t\mathcal{O}\mathcal{O})\mathcal{P} = {}_t\mathcal{P}id\mathcal{P} = {}_t\mathcal{P}\mathcal{P} = id$. Hence $\mathcal{O}\mathcal{P}$ are unitary by proposition 7.3.3.

[3] Let $\mathcal{O} = diag(x_1, \cdots, x_\ell)$ be a real unitary matrix; by definition ${}_t\mathcal{O}\mathcal{O} = I_\ell$. Hence $x_i^2 = 1, \forall i = 1, \cdots, \ell$. So $x_i = \pm 1 \ \forall \ i$.

[4] By virtue of proposition 7.3.3, $\mathcal{O}^2 \circ {}_t\mathcal{O}^2 = \mathcal{O}(\mathcal{O} \circ {}_t\mathcal{O})_t\mathcal{O} = \mathcal{O} \circ id \circ {}_t\mathcal{O} = \mathcal{O} \circ {}_t\mathcal{O} = id$. So $(\mathcal{O}^2 \circ {}_t\mathcal{O}^2)\mathcal{B} = \mathcal{B}$.

Exercises of § 7-4 (page 125)

[1] For any complex unitary operators $\mathcal{O}, \mathcal{P}$, we have ${}_*(\mathcal{O}\mathcal{P}) \circ (\mathcal{O}\mathcal{P}) = {}_*\mathcal{P}_*\mathcal{O} \circ \mathcal{O}\mathcal{P} = {}_*\mathcal{P}\mathcal{P} = id$. So our assertion holds.

[2] Because $\mathcal{O}_*\mathcal{O} = id$ by proposition 7.4.3, we have $\mathcal{O}^m \circ {}_*\mathcal{O}^m = id$, so that $(\mathcal{O}^m \circ {}_*\mathcal{O}^m)\mathcal{B} = \mathcal{B}$.

[3] Let $\mathcal{O} = diag(x_1, \cdots, x_\ell)$ be a complex unitary matrix. Since ${}_*\mathcal{O} = {}_t\bar{\mathcal{O}} = diag(\bar{x_1}, \cdots, \bar{x_\ell})$, we have $id = \mathcal{O} \circ {}_*\mathcal{O} = diag(|x_1|^2, \cdots, |x_\ell|^2)$. Hence $|x_j| = 1\ \forall j = 1, \cdots, \ell$, so that $x_j = e^{i\theta_j}$ for some $\theta_j \in [0, 2\pi)\ \forall j = 1, \cdots, \ell$.

[4] Obvious by definition 7.4.1 and proposition 7.4.2 (ii).

Exercises of § 8-1 (page 131)

[1] (i) $\begin{vmatrix} \mathbb{X}-3 & 2 \\ -2 & \mathbb{X}+3 \end{vmatrix} = \mathbb{X}^2 - 5$. So the eigenvalues are $\pm\sqrt{5}$ over $\mathbb{R}$ and $\mathbb{C}$. Hence the eigenvectors over $\mathbb{R}$ and $\mathbb{C}$ are $(x, \frac{3-\sqrt{5}}{2}x)\ \forall x (\neq 0) \in \mathbb{R}$ and $\mathbb{C}$ respectively.

(ii) $\begin{vmatrix} \mathbb{X}+1 & -2 \\ 2 & \mathbb{X}-2 \end{vmatrix} = \mathbb{X}^2 - \mathbb{X} + 2 = 0$ implies $\mathbb{X} = \frac{1 \pm \sqrt{7}i}{2}$, so that $\nexists$ eigenvalue over $\mathbb{R}$ but $\exists$ eigenvalues $\frac{1 \pm \sqrt{7}i}{2}$ over $\mathbb{C}$.

Hence the eigenvalues and eigenvectors exist over $\mathbb{C}$; $(x, \frac{3+\sqrt{7}i}{4}x)$ $\forall x(\neq 0) \in \mathbb{C}$ are eigenvectors.

(iii) similar to (ii).

[2] Put $\mathbb{X} = (1, \cdots, 1)_{1\times\ell}$; we then have $M_t\mathbb{X} = n_t\mathbb{X}$. So n is an eigenvalue of M.

[3] is also similar to [1](ii).

[4] $\begin{vmatrix} \mathbb{X} - cos\theta & sin\theta \\ -sin\theta & \mathbb{X} - cos\theta \end{vmatrix} = \mathbb{X}^2 - (2cos\theta)\mathbb{X} + 1 = P_M(\mathbb{X})$.
But $\nexists$ root of $P_M(\mathbb{X})$ over $\mathbb{R}$ under the condition. So $\nexists$ eigenvalue of M by proposition 8.1.10. Hence $\nexists$ eigenvector in $\mathbb{R}^2$.

[5] $\frac{d}{dx}e^{\alpha x} = \alpha e^{\alpha x}$, so that $e^{\alpha x}$ is an eigenvector.

Exercises of § 8-2 (page 135)

[1] $V := \mathbb{R}^\ell$ is a finite dimensional vector space over $\mathbb{R}$ with the usual scalar product as a positive definite scalar product. So the spectral basis theorem proves [1].

[2] $_t(u^{-1})(u^{-1}) = (_tu^{-1})(u^{-1}) = (u\ _tu)^{-1} = I^{-1} = I$ holds since u is real unitary.

[3] We have eigenvalues $\varepsilon = \frac{5\pm\sqrt{5}}{2}$ by the equation

$$\begin{vmatrix} X-2 & -1 \\ -1 & X-3 \end{vmatrix} = X^2 - 5X + 5.$$

Hence each case has an eigenvector

$$(1, \varepsilon - 2) = (1, \frac{5\pm\sqrt{5}}{2} - 2) = (1, \frac{1\pm\sqrt{5}}{2}).$$

Hence $\{(1, \frac{1+\sqrt{5}}{2}), (1, \frac{1-\sqrt{5}}{2})\}$ is an orthogonal basis.

Exercises of § 8-3 (page 138)

[1] We know $({}_*u)^{-1} = {}_*(u^{-1})$ since $({}_*u)_*(u^{-1}) = {}_*(u^{-1}u) = id$. Now ${}_*(u^{-1})u^{-1} = ({}_*u)^{-1}u^{-1} = (u_*u)^{-1} = id$, so that u^{-1} is also complex unitary.

[2] If we put $V := \mathbb{C}^\ell$ and $M =: A^{\mathcal{B}}_{\mathcal{B}}(\mathcal{O})$ as in the proof of proposition 8.1.6, we have an orthonormal basis by virtue of proposition 8.3.3.

[3] (i) Let v be an eigenvector of ${}_*\mathcal{O}$ and let S be the linear span of $\{v, \mathcal{O}v, \mathcal{O}^2v, \cdots\}$; further we let w be an eigenvector of $\mathcal{O}|_S : S \longrightarrow S$.

Next we put $S' := \mathbb{C}w$ which is invariant under both $\mathcal{O}$ and ${}_*\mathcal{O}$; then also $\mathcal{O}(S')^\perp \subset (S')^\perp$ holds.

So if we use induction similar to that of the proof in proposition 8.2.3, we might prove easily the following theorem: (Spectral basis theorem for a normal operator)

For any normal operator $\mathcal{O}$, we have an orthonormal basis of V consisting of eigenvectors of $\mathcal{O}$, where V is a finite dimensional vector space over $\mathbb{C}$ with a positive definite hermitian form.

(ii) Since ${}_*\mathcal{P}$ is also normal, we have only to show that $\mathcal{O}_*\mathcal{P}$ is normal. But it is easy to verify $(\mathcal{O}_*\mathcal{P})_*(\mathcal{O}_*\mathcal{P}) = {}_*(\mathcal{O}_*\mathcal{P})(\mathcal{O}_*\mathcal{P})$.

[4] Because ${}_{**}\mathcal{O} = \mathcal{O}$, it is evident.

Exercises of § 8-4 (page 141)

[1] Let a complex unitary operator $\mathcal{O}$ has an eigenvector v with its eigenvalue ε. But then $\|\mathcal{O}v\| = \|v\| = |\varepsilon|\|v\| \neq 0$. $\therefore$ ε is nonzero.

[2] Put $\mathcal{O} := \begin{pmatrix} i & i \\ i & -i \end{pmatrix}$; then ${}_*\mathcal{O} = {}_t\bar{\mathcal{O}} = \begin{pmatrix} -i & -i \\ -i & i \end{pmatrix}$. $\mathcal{O}_*\mathcal{O} \neq id$, so $\mathcal{O}$ is not a complex unitary matrix.

[3] Put $\mathcal{O} := diag(e^{i\theta_1}, \cdots, e^{i\theta_n})$ $\forall\theta_i \in \mathbb{R}$; then ${}_*\mathcal{O} = {}_t\bar{\mathcal{O}} = diag(e^{-i\theta_1}, \cdots, e^{-i\theta_n})$.

Now $\mathcal{O}_*\mathcal{O} = I_n$, so $\mathcal{O}$ is a complex unitary matrix.

[4] $\{U_i \mid \forall i = 1, \cdots, n\}$ is the spectral basis of the given unitary matrix.

Exercises of § 8-5 (page 145)

[1] By corollary 8.5.2, any real unitary operator acts on each component of the decomposition of proposition 8.5.1 as either an identity or a reflection or a rotation. So there exists no zero eigenvalue for any real unitary operator.

[2] ${}_t\begin{pmatrix} cos\theta & sin\theta \\ -cos\theta & sin\theta \end{pmatrix}\begin{pmatrix} cos\theta & sin\theta \\ -cos\theta & sin\theta \end{pmatrix} \neq \begin{pmatrix} 1 & 0 \\ 0 & 1 \end{pmatrix}$, so that $\begin{pmatrix} cos\theta & sin\theta \\ -cos\theta & sin\theta \end{pmatrix}$ is not real unitary.

[3] Since $\begin{pmatrix} cos\theta & sin\theta \\ -sin\theta & cos\theta \end{pmatrix}$ denotes a rotation, we may have the eigenvalues $\{\pm 1\}$ for $\theta = m\pi, m = 0, \pm 1, \pm 2, \cdots$

SKETCH OF ANSWERS TO EXERCISES

Exercises of § 9-1 (page 148)

[1] By proposition 9.1.2, $(pq)(M) = p(M)q(M)$. So $c_j = \sum_{i=0}^{j} a_i b_{j-i}$ holds.

[2] It is straightforward since $(pq)(\mathcal{O}) = p(\mathcal{O})q(\mathcal{O})$.

[3] $\forall n \in \mathbb{Z}^{\geq 0}$, $(\mathcal{P}^{-1}\mathcal{O}\mathcal{P})^n = \underbrace{(\mathcal{P}^{-1}\mathcal{O}\mathcal{P})\cdots(\mathcal{P}^{-1}\mathcal{O}\mathcal{P})}_{n-times}$,

$$(\mathcal{P}^{-1}\mathcal{O})\mathcal{O}\cdots\mathcal{O}(\mathcal{O}\mathcal{P}) = \mathcal{P}^{-1}\mathcal{O}^n\mathcal{P}.$$

So $p(\mathcal{P}^{-1}\mathcal{O}\mathcal{P}) = \mathcal{P}^{-1}p(\mathcal{O})\mathcal{P}$ $\forall$ polynomial $p(\mathbb{X}) = F[\mathbb{X}]$.

Exercises of § 9-2 (page 152)

[1] If we let $\mathcal{B} := \{v_1, \cdots, v_\ell\}$, we then have $A^{\mathcal{B}}_{\mathcal{B}}(\mathcal{O}) = diag(\varepsilon_1, \cdots, \varepsilon_\ell) = M$. So $p_M(\mathbb{X}) = |\mathbb{X}I_\ell - M| = (\mathbb{X} - \varepsilon_1)\cdots(\mathbb{X} - \varepsilon_\ell)$.

[2] By Hamilton-Cayley's theorem, $p_M(M) = 0$. Namely $p_M(A^{\mathcal{B}}_{\mathcal{B}}(\mathcal{O})) = 0$. But according to the problem [3] of § 9-1, $p_M(A^{\mathcal{B}'}_{\mathcal{B}'}(\mathcal{O})) = 0$ $\forall$ basis $\mathcal{B}'$ of V. So $p_M(\mathcal{O}) = 0$. Note

Typeset by $\mathcal{A}\mathcal{M}\mathcal{S}$-TEX

that $A_{\mathcal{B}'}^{\mathcal{B}'}(\mathcal{O}) = \mathcal{P}^{-1}A_{\mathcal{B}}^{\mathcal{B}}(\mathcal{O})\mathcal{P}$ for some invertible matrix $\mathcal{P}$.

[3] From the equation $p_M(\mathbb{X}) = 0$, we have eigenvalues ± 1 of M. Let e_1, e_2 be the corresponding eigenvectors to ± 1, respectively. So $\{e_1, e_2\} =: \mathcal{B}'$ may be a basis of $\mathbb{F}^2$.

Let $\mathcal{B}$ be the standard basis of $\mathbb{F}^2$; then $M = A_{\mathcal{B}}^{\mathcal{B}}(\mathcal{O})$ for some operator $\mathcal{O} : \mathbb{F}^2 \longrightarrow \mathbb{F}^2$. However $A_{\mathcal{B}'}^{\mathcal{B}'}(\mathcal{O}) = diag(+1, -1)$, so that M becomes similar to a diagonal matrix.

[4] If $p_M(\mathbb{X}) = (\mathbb{X} - \varepsilon_1)^{i_1} \cdots (\mathbb{X} - \varepsilon_t)^{i_t}$ for distinct diagonal entries $\varepsilon_1, \cdots, \varepsilon_t \in F$, then we have the minimal polynomial $q_M(\mathbb{X})$ of the given diagonal matrix M like $q_M(\mathbb{X}) = (\mathbb{X} - \varepsilon_1) \cdots (\mathbb{X} - \varepsilon_t)$.

Exercises of § 9-3 (page 156)

[1] (a) $810 = 3 \times 254 + 48, 254 = 5 \times 48 + 14, 48 = 3 \times 14 + 6, 14 = 2 \times 6 + 2, 6 = 3 \times 2$. So g.c.d (810,254)=2 and so $2 = 118 \times 254 - 37 \times 810$.

(b),(c) are similar to (a).

[2] (a) $x^2 + 1 = (x-1)(x+1) + 2, x + 1 = \frac{1}{2}(x+1)2$. So g.c.d $(x^2+1, x+1) = 1$ and hence $1 = \frac{1}{2}(x^2+1) - \frac{1}{2}(x-1)(x+1)$.

(b),(c) are similar to (a).

[3] If we put $a_m = b_n = 1$ in the proof of the proposition

9.3.1, then all the polynomials appearing therein have integer coefficients.

[4] Let $p(\mathbb{X}) = a_m\mathbb{X}^m + a_{m-1}\mathbb{X}^{m-1} + \cdots + a_1\mathbb{X} + a_0$. We then have $p(\mathbb{X}) = a_m(\mathbb{X}^m + \frac{a_{m-1}}{a_m}\mathbb{X}^{m-1} + \cdots + \frac{a_0}{a_m})$, where we assume $a_m \neq 0$. Next we put $\widetilde{p}(\mathbb{X}) = \mathbb{X}^m + \frac{a_{m-1}}{a_m}\mathbb{X}^{m-1} + \cdots + \frac{a_0}{a_m}$ and apply corollary 9.3.3 to $\widetilde{p}(\mathbb{X})$.

We then have $\widetilde{p}(\mathbb{X}) = (\mathbb{X} - r_1)\cdots(\mathbb{X} - r_m)$ by induction. Hence $p(\mathbb{X}) = a_m(\mathbb{X} - r_1)\cdots(\mathbb{X} - r_m)$ for some $r_1, \cdots, r_m \in F$.

[5] $p(1) = 0$, so $p(\mathbb{X}) = (\mathbb{X} - 1)(\mathbb{X}^2 - 5\mathbb{X} + 6)$. Put $q(\mathbb{X}) = \mathbb{X}^2 - 5\mathbb{X} + 6$; then $q(2) = 0$, so $q(\mathbb{X}) = (\mathbb{X} - 2)(\mathbb{X} - 3)$.

$\therefore\ p(\mathbb{X}) = (\mathbb{X} - 1)(\mathbb{X} - 2)(\mathbb{X} - 3)$.

Exercises of § 9-4 (page 159)

[1] $\mathbb{X}^7 - \mathbb{X} = \mathbb{X}(\mathbb{X}^3 + 1)(\mathbb{X} - 1)(\mathbb{X}^2 + \mathbb{X} + 1)$, so divisors are as follows: $\mathbb{X}, \mathbb{X}^3 + 1, \mathbb{X} - 1, \mathbb{X}^2 + \mathbb{X} + 1, \mathbb{X}(\mathbb{X}^3 + 1), \mathbb{X}(\mathbb{X} - 1), \mathbb{X}(\mathbb{X}^2 + \mathbb{X} + 1), (\mathbb{X}^3 + 1)(\mathbb{X} - 1), (\mathbb{X}^3 + 1)(\mathbb{X}^2 + \mathbb{X} + 1), (\mathbb{X} - 1)(\mathbb{X}^2 + \mathbb{X} + 1), \mathbb{X}(\mathbb{X}^3 + 1)(\mathbb{X} - 1), \mathbb{X}(\mathbb{X}^3 + 1)(\mathbb{X}^2 + \mathbb{X} + 1), \mathbb{X}(\mathbb{X} - 1)(\mathbb{X}^2 + \mathbb{X} + 1), (\mathbb{X}^3 + 1)(\mathbb{X} - 1)(\mathbb{X}^2 + \mathbb{X} + 1)$.

[2] g.c.d.$(\mathbb{X} + 1, \mathbb{X}^2 - 1) = \mathbb{X} + 1$ according to proposition 9.4.3.

[3] By corollary 9.3.3.,$(\mathbb{X} + 1)|(\mathbb{X}^m + 1)$ holds if m is odd.

[4] Put $p(\mathbb{X}) = \mathbb{X}^m - 1$; then $p(1) = 0$. So by corollary 9.3.3., $(\mathbb{X} - 1)|(\mathbb{X}^m - 1)\ \forall m \in \mathbb{N}$.

[5] $F[\mathbb{X}]$.

Exercises of § 9-5 (page 162)

[1] $p(\mathbb{X}) = (f_0 + f_1\mathbb{X})(f_0' + f_1'\mathbb{X} + f_2'\mathbb{X}^2)$ for some $f_0, f_1, f_0', f_1', f_2' \in F$, where $f_1, f_2' \neq 0$. But then $r = -\dfrac{f_0}{f_1}$ is a root of $p(\mathbb{X})$ in F.

[2] (i) Since $(p+q)(\mathbb{X}) = p(\mathbb{X}) + q(\mathbb{X})$ and $(pq)(\mathbb{X}) = p(\mathbb{X}) \cdot q(\mathbb{X})$, we have that $(p+q)'(\mathbb{X}) = (p'+q')(\mathbb{X})$ and $(pq)'(\mathbb{X}) = (p'q)(\mathbb{X}) + (pq')(\mathbb{X})$.

(ii)($\Rightarrow$) $p(\mathbb{X}) := (\mathbb{X}-r)^m(f_0 + f_1\mathbb{X} + \cdots + f_\ell\mathbb{X}^\ell), m > 1$ $\Rightarrow p'(\mathbb{X}) = m(\mathbb{X}-r)^{m-1}(f_0 + f_1\mathbb{X} + \cdots + f_\ell\mathbb{X}^\ell) + (\mathbb{X} - r)^m(f_1 + 2f_2\mathbb{X} + \cdots + \ell f_\ell \mathbb{X}^{\ell-1})$. So $p'(r) = 0$.

($\Leftarrow$) We define $\int p'(\mathbb{X})d\mathbb{X} := p(\mathbb{X}) + c$, where c is any constant; then $\dfrac{d}{dx}\int p'(\mathbb{X})d\mathbb{X} = p'(\mathbb{X})$. Since $p'(r) = 0$, we have $p'(\mathbb{X}) = (\mathbb{X}-r)(g_0 + g_1\mathbb{X} + \cdots + g_\ell\mathbb{X}^\ell)$ for some polynomial $g_0 + g_1\mathbb{X} + \cdots + g_\ell\mathbb{X}^\ell$. So $p(\mathbb{X}) + c = \int p'(\mathbb{X})d\mathbb{X} = \int(\mathbb{X}-r)(g_0 + g_1\mathbb{X} + \cdots + g_\ell\mathbb{X}^\ell)d\mathbb{X} = (\mathbb{X}-r)^2q(\mathbb{X}) + d$ for some $d \in F$ and for some $q(\mathbb{X}) \in F[\mathbb{X}]$. But then $c = d$ since $p(r) = 0$. $\therefore$ $p(\mathbb{X}) = (\mathbb{X}-r)^2q(\mathbb{X})$.

Exercises of § 10-1 (page 166)

[1] $\forall v \in Ker\, p(\mathcal{O}), f \cdot p(\mathcal{O})(v) = f\{p(\mathcal{O})v\} = 0$. So $Ker\, p(\mathcal{O}) \subset Ker\ f \cdot p(\mathcal{O})$. Next $\forall v \in Ker\ f \cdot p(\mathcal{O})$, $\{f \cdot p(\mathcal{O})\}v = 0 = f\{p(\mathcal{O})(v)\}$.

So $p(\mathcal{O})(v) = 0$, so that $v \in Ker\ p(\mathcal{O})$. Hence $Ker\ f \cdot p(\mathcal{O}) \subset Ker\ p(\mathcal{O})$. $\therefore$ $Ker\ f \cdot p(\mathcal{O}) = Ker\ p(\mathcal{O})$.

[2] $\forall w \in Im\ p_2(\mathcal{O}), \exists v \in V$ such that $p_2(\mathcal{O})v = w$.

$$p_1(\mathcal{O})w = p_1(\mathcal{O})p_2(\mathcal{O})v = p(\mathcal{O})v = 0.$$

So $w \in Ker\ p_1(\mathcal{O})$. $\therefore$ $Im\ p_2(\mathcal{O}) \subset Ker\ p_1(\mathcal{O})$.

Conversely, let $v \in Ker\ p_1(\mathcal{O})$. Since $1 = q_1p_1 + q_2p_2$ for some $q_1, q_2 \in F[\mathbb{X}]$, we have

$$\begin{aligned} v &= q_1(\mathcal{O})p_1(\mathcal{O})v + q_2(\mathcal{O})p_2(\mathcal{O})v \\ &= q_2(\mathcal{O})p_2(\mathcal{O})v \\ &= p_2(\mathcal{O})q_2(\mathcal{O})v. \end{aligned}$$

So $v \in Im\ p_2(\mathcal{O})$, i.e., $Im\ p_2(\mathcal{O}) \supset Ker\ p_1(\mathcal{O})$.

Hence $Im\ p_2(\mathcal{O}) = Ker\ p_1(\mathcal{O})$.

[3] By corollary 10.1.4., $V = Ker(\mathcal{O} - f_1I) \oplus \cdots \oplus Ker(\mathcal{O} - f_\ell I)$. Also by corollary 9.2.2 and proposition 8.1.10, $Ker\ (\mathcal{O} - f_iI) \neq \{0\}$ and f_i is an eigenvalue of $\mathcal{O}$ $\forall i = 1, \cdots, \ell$. So V has a basis consisting of eigenvectors of $\mathcal{O}$.

Exercises of § 10-2 (page 169)

[1] By virtue of proposition 10.2.4, $\mathcal{O} = f \cdot I$ for some $f \in F$. So any nonzero vector of V becomes an eigenvector with an eigenvalue f. So our assertion holds.

[2] $\{0\}$, $\mathbb{R} \times \{0\}$, $\mathbb{R}^2 = \mathbb{R} \times \mathbb{R}$.

[3] Let F_v be the collection of all subspaces with a basis of V; then $X =< F_v, \subset>$ becomes an ordered set. Any totally ordered subset of X has an upper bound in X obviously considering the union operation U.

So by the Zorn's lemma, we have a maximal element of X, which is nothing but V itself. Similarly it holds for any $\widetilde{S}$-invariant subspace of V.

[4] Since any $\widetilde{S}$-invariant subspaces of V has a basis, we may say that $\forall v_1, v_2 \in V, \exists L \in LM(V,V)$ such that $Lv_1 = v_2$ for $v_1 \neq 0$. Hence $\{0\}$ and V are $\widetilde{S}$-invariant subspaces of V only.

Exercises of § 10-3 (page 172)

[1] With respect to the basis $\mathcal{B} = \{v, (\mathcal{O} - fI)v, \cdots, (\mathcal{O} - fI)^{s-1}v\}$ as in proposition 10.3.4, we have

$$A_{\mathcal{B}}^{\mathcal{B}}(\mathcal{O}) = diag(f, \cdots, f) + \begin{pmatrix} 0 & 0 & \cdots & \cdots & 0 \\ 1 & \ddots & \ddots & \ddots & \vdots \\ \vdots & \ddots & \ddots & \ddots & \vdots \\ \vdots & & \ddots & \ddots & 0 \\ 0 & \cdots & \cdots & 1 & 0 \end{pmatrix}_{s \times s}.$$

For some operator D and N, we have

$$A_{\mathcal{B}}^{\mathcal{B}}(D) = diag(f, \cdots, f)$$

$$\text{and } A^{\mathcal{B}}_{\mathcal{B}}(N) = \begin{pmatrix} 0 & 0 & \cdots & \cdots & 0 \\ 1 & \ddots & \ddots & \ddots & \vdots \\ \vdots & \ddots & \ddots & \ddots & \vdots \\ \vdots & & \ddots & \ddots & 0 \\ 0 & \cdots & \cdots & 1 & 0 \end{pmatrix}_{s \times s} \text{ respectively.}$$

Now $A^{\mathcal{B}}_{\mathcal{B}}(D) \cdot A^{\mathcal{B}}_{\mathcal{B}}(N) = A^{\mathcal{B}}_{\mathcal{B}}(N) \cdot A^{\mathcal{B}}_{\mathcal{B}}(D)$ shows $DN = ND$,while $\{A^{\mathcal{B}}_{\mathcal{B}}(N)\}^s = A^{\mathcal{B}}_{\mathcal{B}}(N^s) = 0$ shows $N^s = 0$.

[2] According to proposition 10.3.4, $\mathcal{O}$ can have f as its eigenvalue and nothing else.

[3] By virtue of definition 8.1.1, we have an eigenvector $v \in V$ such that $\mathcal{O}v = \varepsilon v$. So $(\mathcal{O} - \varepsilon I)v = 0$ but $(\mathcal{O} - \varepsilon I)^0 v = Iv = v \neq 0$. Hence v becomes $(\mathcal{O} - \varepsilon I)$-cyclic with period 1.

[4] Since any vector $w \in V$ is a linear combination of $v, (\mathcal{O} - fI)v, \cdots, (\mathcal{O} - fI)^{s-1}v$, we see that w becomes $(\mathcal{O} - fI)$-cyclic. Here we assumed that V is generated by $v, (\mathcal{O} - fI)v, \cdots, (\mathcal{O} - fI)^{s-1}v$.

Exercises of § 10-4 (page 176)

[1] Let M be a real symmetric matrix; then M is similar to a diagonal matrix $D = diag(\varepsilon_1, \cdots, \varepsilon_\ell)$, where $\varepsilon_1, \cdots, \varepsilon_\ell$ are real eigenvalues of M, by the spectral basis theorem.

Hence $p_M(\mathbb{X}) = p_D(\mathbb{X}) = (\mathbb{X} - \varepsilon_1) \cdots (\mathbb{X} - \varepsilon_\ell)$, which is similarly applied to a hermitian matrix.

[2] By virtue of the spectral basis theorem in § 8-4, this is also straightforward.

[3] (i) Let $V = V_1 \oplus \cdots \oplus V_\ell$ be the Jordan decomposition consisting of $\mathcal{O}$-invariant cyclic subspaces V_i. By exercise [1] of § 10-3, we get $\mathcal{O}|_{V_i} = D|_{V_i} + N|_{V_i}$. Now put $D := \sum_{i=1}^{\ell} D|_{V_i}$ and $N := \sum_{i=1}^{\ell} N|_{V_i}$.

(ii) If $A_{\mathcal{B}}^{\mathcal{B}}(\mathcal{O}) = \begin{pmatrix} f & 1 & \cdots & \cdots & 0 \\ 0 & \ddots & \ddots & \ddots & \vdots \\ \vdots & \ddots & \ddots & \ddots & \vdots \\ \vdots & & \ddots & \ddots & 1 \\ 0 & \cdots & \cdots & 0 & f \end{pmatrix}$ as in proposition 10.3.4, then $F_t(1, 0, \cdots, 0)$ is the only eigenspace of $\mathcal{O}$.

[4] (i) Put $v_1 := {}_t(1,0,0,0), v_2 := {}_t(0,1,0,0), v_3 := {}_t(0,0,1,1)$; then v_1, v_2, v_3 are $(\mathcal{O}-I), (\mathcal{O}-I), (\mathcal{O}-2I)$-cyclic vectors with the periods 1,1,2 respectively. Now put

$V_1 :=$ the subspace of V generated by $\{v_1\}$,

$V_2 :=$ the subspace of V generated by $\{v_2\}$,

$V_3 :=$ the subspace of V generated by $\{v_3, \mathcal{O}v_3\}$.

We then have $V = F^4 = V_1 \oplus V_2 \oplus V_3$.

(ii)~(v) are similar to (i).

[5] Let $\mathcal{B} = \{v_1, \cdots, v_\ell\}$ be a fan basis of $V = S_\ell$. Since $S_i\ \forall i = 1, \cdots, \ell$ becomes $\mathcal{O}$-invariant, we have the form

$A_{\mathcal{B}}^{\mathcal{B}}(\mathcal{O}) = \begin{pmatrix} * & \cdots & \cdots & \cdots & * \\ 0 & \ddots & \ddots & \ddots & \vdots \\ \vdots & \ddots & \ddots & \ddots & \vdots \\ \vdots & & \ddots & \ddots & \vdots \\ 0 & \cdots & \cdots & 0 & * \end{pmatrix}$, which is upper triangular.

Next let $V = V_1 \oplus \cdots \oplus V_\ell$ be the Jordan decomposition and let $\mathcal{B}_j := \{(\mathcal{O} - f_j I)^{s_j - 1} v_j, \cdots, (\mathcal{O} - f_j I) v_j, v_j\}$ be the Jordan basis of each V_j relative to $\mathcal{O}$. Let V_{j_k} be the subspace generated by $(\mathcal{O} - f_j I)^{s_j - 1} v_j, \cdots, (\mathcal{O} - f_j I)^{s_j - k} v_j$ for $k = 1, \cdots, s_j$.

Obviously $\{V_{j_1}, V_{j_2}, \cdots, V_{j_{s_j}}\}$ becomes a fan of $\mathcal{O}|_{V_j}$. So $\mathcal{B}_j$ is a fan basis of V_j $\forall\, j$.

Hence $\bigcup_{j=1}^{\ell} \mathcal{B}_j$ becomes a fan basis of V.